MW01482645

hands-on
science
An Inquiry Approach

Land, Water, and Sky
for Grades K–2

Senior Author

Jennifer Lawson

PORTAGE &
MAIN PRESS

Winnipeg • Manitoba • Canada

© 2019 Jennifer Lawson

Pages of this publication designated as reproducible with the following icon ![icon] may be reproduced under licence from Access Copyright. All other pages may be reproduced only with the permission of Portage & Main Press, or as permitted by law.

All rights are otherwise reserved, and no part of this publication may be reproduced, stored in a retrieval system, or transmitted in any form or by any means—electronic, mechanical, photocopying, scanning, recording, or otherwise—except as specifically authorized.

Portage & Main Press gratefully acknowledges the financial support of the Province of Manitoba through the Department of Sport, Culture, and Heritage and the Manitoba Book Publishing Tax Credit, and the Government of Canada through the Canada Book Fund (CBF), for our publishing activities.

Hands-On Science for British Columbia
Land, Water, and Sky for Grades K–2
An Inquiry Approach

ISBN: 978-1-55379-797-5
Printed and bound in Canada by Prolific Group

1 2 3 4 5 6 7 8 9 10 11

Download the image banks and reproducibles that accompany this book by going to the Portage & Main Press website at <www.portageandmainpress.com/product/HOSLANDWATERSKYK2/>. Use the password **WATERCYCLE** to access this free download. For step-by-step instructions to access this download, see the Appendix, page 165.

The publisher has made every effort to acknowledge all sources of photographs used in the image banks and to ensure the authenticity of all Indigenous resources. The publisher would be grateful if any errors or omissions were pointed out, so that they may be corrected.

A special thank-you to the following people for their generous contributions to this project:

Project Consultants:
Faye Brownlie
Kathleen Gregory

Science Consultant:
Rosalind Poon

Early Years Consultants:
Lisa Schwartz
Deidre Sagert

Indigenous Consultants:
Brett D. Huson, Gitxsan
Melanie Nelson, Stó:lō and In-SHUCK-ch

Makerspace Contributors:
Joan Badger
Todd Johnson

Curriculum Correlation Consultant:
Susan Atcheson

Book and Cover Design:
Relish New Brand Experience Inc.

Cover Photo:
Adobestock

Illustrations:
ArtPlus Ltd.
26 Projects
Jess Dixon

www.portageandmainpress.com
Winnipeg, Manitoba
Treaty 1 Territory and homeland
of the Métis Nation

MIX
Paper from responsible sources
FSC® C006215

Contents

Portage & Main Press, 2019 · Hands-On Science for British Columbia · Land, Water, and Sky for Grades K–2 · ISBN: 978-1-55379-797-5

Introduction to *Hands-On Science*

About *Hands-On Science*

Hands-On Science helps develop students' scientific literacy through active inquiry, problem solving, and decision making. With each activity in *Hands-On Science*, students are encouraged to explore, investigate, and ask questions as a means of heightening their own curiosity about the world around them. Students solve problems through firsthand experiences and by observing and examining objects within their environment. In order for young students to develop scientific literacy, concrete experience is of utmost importance—in fact, it is essential.

Format of *Hands-On Science*

The redesigned Science Curriculum for British Columbia (<https://curriculum.gov.bc.ca/>) is based on a "**Know-Do-Understand**" model. The three elements—Content (Know), Curricular Competencies (Do), and Big Ideas (Understand) all work together to support deeper learning. *Hands-On Science* promotes this model through its inquiry-based, student-centred approach. As such, it is structured around the following elements.

The **Big Ideas** are broad concepts introduced in kindergarten and expanded upon in subsequent grades, fostering a deep understanding of science. The Big Ideas form the basis of the *Hands-On Science* modules to address important concepts in biology, chemistry, physics, and earth/space science.

The **Core Competencies** are embedded throughout the curriculum and throughout *Hands-On Science*. These competencies enable students to engage in deeper lifelong learning.

Core Competencies

Thinking	■ knowledge, skills, and processes that enable students to explore problems, weigh alternatives, and arrive at solutions
	■ problem solving and making effective decisions, and applying them to real-world contexts
Communication	■ effectively reading, writing, speaking, listening, viewing, and representing
	■ using a variety of information sources and digital tools
Personal and Social	■ relates to a student's identity as an individual and as a member of a group or community
	■ contributing to the care of themselves, others, and the larger community

The **Learning Standards** are made up of **Curricular Competencies** and **Content**. **Curricular Competencies** are skills, strategies, and processes students develop as they explore science through hands-on activities. Curricular Competencies are addressed further on page 33.

The **Content** of the Science Curriculum for British Columbia and *Hands-On Science* is concept-based and relates directly to the Big Ideas. The Content relies on cross-cutting concepts developed throughout the grade levels, including:

■ cause and effect
■ change
■ cycles
■ evolution
■ form and function
■ interactions
■ matter and energy
■ order
■ patterns
■ systems

▶

Portage & Main Press, 2019 · *Hands-On Science for British Columbia · Land, Water, and Sky for Grades K–2* · ISBN: 978-1-55379-797-5

The Multi-Age Approach

Hands-On Science is designed with a multi-age approach to meet the needs of students in kindergarten to grade two (K–2). Each module explores the Big Ideas, Core Competencies, and Learning Standards for K–2. This approach provides teachers and students with flexible, personalized learning opportunities.

Inquiry and Science

Throughout **Hands-On Science**, as students explore science concepts, they are encouraged to ask questions to guide their own learning. The inquiry model is based on five components:

1. formulating questions
2. gathering and organizing information, evidence, or data
3. interpreting and analyzing information, evidence, or data
4. evaluating information, evidence, or data, and drawing conclusions
5. communicating findings

Using this model, teachers facilitate the learning, and students drive the process through inquiry. As such, the approach focuses on students' self-reflections as they ask questions, discover answers, and communicate their understanding. An inquiry approach begins with structured inquiry, moves to guided inquiry and, finally, results in open inquiry.

Structured Inquiry	■ The teacher provides the initial question and structures the procedures to answer it. ■ Students follow the given procedures and draw conclusions to answer the given question.
Guided Inquiry	■ The teacher provides the initial question. ■ Students are involved in designing ways to answer the question and communicate their findings.
Open Inquiry	■ Students formulate their own question(s), design and follow through with a developed procedure, and communicate their findings and results.

Inquiry takes time to foster and requires scaffolding from a structured approach to more open inquiry as students gain skills and experience.

In **Hands-On Science**, the focus of most activities is on guided inquiry, as teachers pose the main question for the lesson, based on the Learning Standards. Students are involved in generating further inquiry questions to personalize learning, but will continue to benefit from guidance and support from the teacher.

> Open inquiry activities are only successful if students are motivated by intrinsic interests and if they are equipped with the skills to conduct their own research study. (Banchi and Bell, 2008)

The Goals of Science Education in British Columbia

Science plays a fundamental role in the lives of Canadians. The Science Curriculum for British Columbia (<https://curriculum.gov.bc.ca/>) states:

> Science provides opportunities for us to better understand our natural world. Through science, we ask questions and seek answers to grow our collective scientific knowledge. We continually revise and refine our knowledge as we acquire new evidence. While maintaining our respect for evidence, we are aware that our scientific knowledge is provisional and is influenced by our culture, values, and ethics. Linking traditional and contemporary First Peoples understandings and current scientific knowledge enables us to make meaningful connections to our everyday lives and the world beyond.

> The Science curriculum takes a place-based approach to science learning. Students will develop place-based knowledge about the area in which they live, learning about and building on First Peoples knowledge and other traditional knowledge of the area. This provides a basis for an intuitive relationship with and respect for the natural world; connections to their ecosystem

Portage & Main Press, 2019 · Hands-On Science for British Columbia · Land, Water, and Sky for Grades K–2 · ISBN: 978-1-55379-797-5

and community; and a sense of relatedness that encourages lifelong harmony with nature.

The Science Curriculum for British Columbia identifies five goals that form the foundation of science education. In keeping with this focus on scientific literacy, these goals are the bases for the lessons in **Hands-On Science**. The Science Curriculum for British Columbia contributes to students' development as educated citizens through the achievement of the following goals. Students are expected to develop:

1. an understanding and appreciation of the nature of science as an evidence-based way of knowing the natural world that yields descriptions and explanations that are continually being improved within the context of our cultural values and ethics
2. place-based knowledge of the natural world and experience in the local area in which they live by accessing and building on existing understandings, including those of First Peoples
3. a solid foundation of conceptual and procedural knowledge in science that they can use to interpret the natural world and apply to new problems, issues, and events; to further learning; and to their lives
4. the habits of mind associated with science—a sustained curiosity; an appreciation for questions; an openness to new ideas and consideration of alternatives; an appreciation of evidence; an awareness of assumptions and a questioning of given information; a healthy, informed skepticism; a seeking of patterns, connections, and understanding; and a consideration of social, ethical, and environmental implications
5. a lifelong interest in science and the attitudes that will make them scientifically literate citizens who bring a scientific perspective, as appropriate, to social, moral, and ethical

decisions and actions in their own lives, culture, and the environment

Hands-On Science Principles

- Effective science education involves hands-on inquiry, problem solving, and decision making.
- The development of Big Ideas, Core Competencies, Curricular Competencies, and Content form the foundation of science education.
- Children have a natural curiosity about science and the world around them. This curiosity must be maintained, fostered, and enhanced through active learning.
- Science activities must be meaningful, worthwhile, and related to real-life experiences.
- The teacher's role is to facilitate activities and encourage critical thinking and reflection. Children learn best by doing, rather than by just listening. Instead of simply telling, the teacher, therefore, should focus on formulating and asking questions, setting the conditions for students to ask their own questions, and helping students to make sense of the events and phenomena they have experienced.
- Science should be taught in conjunction with other school subjects. Themes and topics of study should integrate ideas and skills from several core areas whenever possible.
- Science education should encompass, and draw on, a wide range of educational resources, including literature, nonfiction research material, audio-visual resources, and technology, as well as people and places in the local community.
- Science education should be infused with knowledge and worldviews of Indigenous peoples, as well as other diverse multicultural perspectives.

Portage & Main Press, 2019 · *Hands-On Science for British Columbia · Land, Water, and Sky for Grades K–2* · ISBN: 978-1-55379-797-5

- Science education should emphasize personalized learning. Personalized learning also focuses on enhancing student engagement and providing them with choices to explore and investigate ideas. Personalized learning also encompasses place-based learning, where learning focuses on the local environment.
- Science education is inclusive in nature. Learning opportunities should meet the diverse needs of all students through differentiated instruction and individualized learning experiences.
- Self-assessment is an integral part of science education. Students should be involved in reflecting on their work and setting new goals based on their reflections which, in turn, enables them to take control of their learning.
- Teacher assessment of student learning in science should be designed to focus on performance and understanding, and should be conducted through meaningful assessment techniques implemented throughout each module.

Cultural Connections

To acknowledge and celebrate the cultural diversity represented in Canadian classrooms, it is important to infuse cultural connections into classroom learning experiences. It is essential for teachers to be aware of the cultural makeup of their class and to celebrate these diverse cultures by making connections to curricular outcomes. In the same way, it is important to explore other cultures represented in the community and beyond, to encourage intercultural understanding and harmony. For example, teachers in British Columbia should make connections to the local cultural communities to highlight their contributions to the province. Throughout **Hands-On Science**, suggestions are made for connecting science topics to cultural explorations and activities.

Portage & Main Press, 2019 · Hands-On Science for British Columbia · Land, Water, and Sky for Grades K–2 · ISBN: 978-1-55379-797-5

Indigenous Perspectives and Knowledge

Indigenous peoples are central to the Canadian context, and it is important to infuse Indigenous knowledge into the learning experiences of all students. The intentional integration of Indigenous knowledge in **Hands-On Science** helps to address the Calls to Action of the Truth and Reconciliation Commission of Canada, particularly the call to "integrate Indigenous knowledge and teaching methods into classrooms" (Action 62) and "build student capacity for intercultural understanding, empathy and mutual respect" (Action 63).

Indigenous peoples have depended on the land since time immemorial. The environment shapes the way of life: geography, vegetation, climate, and natural resources of the land determine the methods used to survive. Because they observe the land and its inhabitants, the environment teaches Indigenous peoples to survive. The land continues to shape Indigenous peoples' way of life today because of their ongoing, deep connection with the land. Cultural practices, stories, languages, and knowledge originate from the land.

The traditional territories of the First Peoples cover the entirety of what is now British Columbia. The worldviews of Indigenous peoples and their approaches and contributions to science are now being acknowledged and incorporated into science education. It is also important to recognize the diversity of Indigenous peoples in British Columbia and to focus on both the traditions and contemporary lives of the Indigenous communities in your area. Contact personnel in your school district—Indigenous consultants and/or those responsible for Indigenous education—to find out what resources (e.g., people, books, videos) are available. Many such resources are also featured in **Hands-On Science**.

NOTE: When implementing place-based learning, many opportunities abound to consider Indigenous perspectives and knowledge. Outdoor learning provides an excellent opportunity to identify the importance of place. For example, use a map of the local area to have students identify where the location is in relation to the school. This will help students develop a stronger image of their community and surrounding area.

It is also important to identify on whose traditional territory the school is located, the traditional territory of the location for the place-based learning, as well as the traditional names for both locations. The following map, "First Nations in British Columbia," from Indigenous Services Canada can be used for this purpose: <https://www.aadnc-aandc.gc.ca/DAM/DAM-INTER-BC/STAGING/texte-text/fnmp_1100100021018_eng.pdf>.

Incorporate land acknowledgment once students have learned on whose territory the school and place-based learning location are located. The following example can be used for guidance:

■ We would like to acknowledge that we are gathered today on the traditional, ancestral, and unceded territory of the _____ people, in the place traditionally known as _____.

When incorporating Indigenous perspectives, it is important to value Traditional Ecological Knowledge (TEK):

> Traditional Ecological Knowledge, or TEK, is the most popular term to denote the vast local knowledge First Peoples have about the natural world found in their traditional environment… TEK is, above all, local knowledge based in people's relationship to place. It is also holistic, not subject to the segmentation of contemporary science. Knowledge about a specific plant may include understanding its life cycle, its spiritual connections, its relationship to the seasons and with other plants and animals in its ecosystem, as well as its uses and its stories. (*Science First Peoples Teacher Resource Guide*)

Indigenous peoples developed technologies and survived on this land for millennia because of

Portage & Main Press, 2019 · Hands-On Science for British Columbia · Land, Water, and Sky for Grades K–2 · ISBN: 978-1-55379-797-5

their knowledge of the land. Indigenous peoples used observation and experimentation to refine technologies, such as building canoes and longhouses and discovering food-preservation techniques. As such, TEK serves as an invaluable resource for students and teachers of science.

Indigenous peoples do not view their knowledges as "science" but, rather, from a more holistic perspective, as is reflected in this quote from Dr. Jolly, Cherokee, and President of the Science Museum of Minnesota:

> When I weave a basket, I talk about the different dyes and how you make them and how the Oklahoma clay that we put on our baskets doesn't permeate the cell walls, it deposits on the outside. It makes a very nice dye but if you cut through the reed you'll see white still on the inside of the reed, whereas if I make a walnut dye and if I use as my mordent, alum and I use as my acid cider, that walnut dye will permeate the cell walls. You cut through the reed and it's brown through and through. Now what I've just described is the difference between osmosis and dialysis. That Western science calls those scientific terms is really wonderful, but it's not scientific terms if you are a basket weaver. Our culture incorporates so much of what people would call scientific knowledge and ways of thinking so naturally that we haven't parsed it out and put it in a book and said this is our science knowledge versus our weaver's knowledge. When I weave a basket I also tell the stories of the spirituality and not just the ways of which I dyed it. A basket weaver is as much a scientist, as an artist, and a spiritual teacher. We'd never think that you'd separate out just the science part, but you can't weave a basket without knowing the science. (*Science First Peoples Teacher Resource Guide*)

Throughout **Hands-On Science**, there are many opportunities to incorporate culturally appropriate teaching methodologies from an Indigenous worldview. First Peoples Pedagogy indicates that making connections to the local community is central to learning (*Science First Peoples Teacher Resource Guide*). As one example, Elders and Knowledge Keepers offer a wealth of knowledge that can be shared with students. Consider inviting a local Elder or Knowledge Keeper as a guest into the classroom in connection with specific topics being studied (as identified within the given lessons throughout the module). An Elder or Knowledge Keeper can guide a nature walk, share stories and experiences, share traditional technologies, and help students understand Indigenous peoples' perspectives of the natural world. Elders and Knowledge Keepers will provide guidance for learners and opportunities to build bridges between the school and the community.

Here are a few suggestions about working with Elders and Knowledge Keepers:

- Elders and Knowledge Keepers have a deep spirituality that influences every aspect of their lives and teachings. They are recognized because they have earned the respect of their community through wisdom, harmony, and balance in their actions and teachings. (see "Aboriginal Elder Definition" at <https://www.ictinc.ca/blog/aboriginal-elder-definition>).
- Some Indigenous keepers of knowledge are more comfortable being called "Knowledge Keepers" than "Elders." Be sensitive to their preference. In many communities, there are also "Junior Elders" who may also be invited to share their knowledge with students and school staff.
- Elders and Knowledge Keepers may wish to speak about what seems appropriate to them, instead of being directed to talk about something specific. It is important to respect this choice and not be directive about what an Elder or Knowledge Keeper will talk about during their visit.

Portage & Main Press, 2019 · Hands-On Science for British Columbia · Land, Water, and Sky for Grades K–2 · ISBN: 978-1-55379-797-5

- It is important to properly acknowledge any visiting Elders or Knowledge Keepers and their knowledge, as they have traditionally been and are recognized within Indigenous communities as highly esteemed individuals. There are certain protocols that should be followed when inviting an Elder or Knowledge Keeper to support student learning in the classroom or on the land. The *Science First Peoples Teacher Resource Guide* offers guidelines and considerations for this.

It is especially important to connect with Indigenous communities, Elders, and Knowledge Keepers in your local area, and to study local issues related to Indigenous peoples in British Columbia. Consider contacting Indigenous education consultants within your local school district or with the British Columbia Ministry of Education to access referrals. The following link provides a province-wide list of Indigenous contacts: <www.bced.gov.bc.ca/apps/imcl/imclWeb/AB.do>. Also, consider contacting local Indigenous organizations for referrals to Elders and Knowledge Keepers. Such organizations may also be able to offer resources and opportunities for field trips and place-based learning.

NOTE: It is important for educators to understand the significant contribution that Elders, Knowledge Keepers, and Indigenous communities make when they share their traditional knowledge. In their culture of reciprocity, this understanding should extend past giving a gift or honorarium to an Elder or Knowledge Keeper for sharing sacred knowledge. As such, educators should think deeply about reciprocity and what they can do beyond inviting Indigenous guests to their classrooms. Educators can expand their own learning and become connected to Indigenous people by, for example, engaging in Indigenous community events, working with the Education Department of the local Nations, or exploring ways to continue developing the relationship between the local Nations and educators in the district.

The First Nations Education Steering Committee of British Columbia has articulated the following **First Peoples Principles of Learning**:

- Learning ultimately supports the well-being of the self, the family, the community, the land, the spirits, and the ancestors.
- Learning is holistic, reflexive, reflective, experiential, and relational (focused on connectedness, on reciprocal relationships, and a sense of place).
- Learning involves recognizing the consequences of one's actions.
- Learning involves generational roles and responsibilities.
- Learning recognizes the role of Indigenous knowledge.
- Learning is embedded in memory, history, and story.
- Learning involves patience and time.
- Learning requires exploration of one's identity.
- Learning involves recognizing that some knowledge is sacred and only shared with permission and/or in certain situations.

These principles generally reflect First Peoples pedagogy, and have been considered in the development of **Hands-On Science**.

The First People Principles of Learning (FPPL) is a framework for approaching learning, or a worldview on what learning is and how it happens. Teachers are encouraged to find their own meaning in them, explore them with their class, and take them up in a way that is meaningful to them. They are embedded in the new curriculum—the new curriculum was created based on these principles. Teachers can make their own connections to the FPPL through the **Hands-On Science** resource. (Melanie Nelson, February 12, 2018)

Portage & Main Press, 2019 · *Hands-On Science for British Columbia · Land, Water, and Sky for Grades K–2* · ISBN: 978-1-55379-797-5

It is also important to note that the *Science First Peoples Teacher Resource Guide* recommends a 7E model for guiding experiential learning activities in science. This model suggests that the following elements are essential to the learning experience:

The 7E Model

Environment	▪ using the local land (place-based learning)
Engage	▪ inspiring curiosity and activating knowledge
Explore	▪ investigating science concepts through hands-on experiences
Elders	▪ connecting local Knowledge Keepers to learning
Explain	▪ describing observations and sharing new knowledge
Elaborate	▪ extending and enhancing learning
Evaluation	▪ providing opportunities for students to demonstrate understanding and skills

These seven elements are strongly evident in the approach used in **Hands-On Science**, as is explained in the following sections.

For more information on First Peoples Pedagogy and First Peoples Principles of Learning, please see the *Science First Peoples Teacher Resource Guide*.

NOTE: Indigenous resources recommended in **Hands-On Science** are considered to be authentic resources, meaning that they reference the Indigenous community they came from, they state the individual who shared the story and gave permission for the story to be used publicly, and the person who originally shared the story is Indigenous. Stories that are works of fiction were written by an Indigenous author. For more information, please see *Authentic First Peoples Resources* at: <www.fnesc.ca/learningfirstpeoples/>.

References

"Aboriginal Contacts—Basic Information." British Columbia Ministry of Education. <www.bced.gov.bc.ca/apps/imcl/imclWeb/AB.do>

Banchi, Heather, and Randi Bell. "The Many Levels of Inquiry." *Science and Children*, 46.2 (2008): 26–29.

British Columbia Ministry of Education. *BC's New Curriculum*. 2016. <https://curriculum.gov.bc.ca/>

"First Nations in British Columbia." Indigenous Services Canada. <https://www.aadnc-aandc.gc.ca/DAM/DAM-INTER-BC/STAGING/texte-text/fnmp_1100100021018_eng.pdf>

"Aboriginal Elder Definition." Indigenous Corporate Training, Inc., 2012. <https://www.ictinc.ca/blog/aboriginal-elder-definition>

"Learning First Peoples Classroom Resources." First Nations Education Steering Committee. <http://www.fnesc.ca/learningfirstpeoples/> (includes *First Peoples Principles of Learning* and *Authentic First Peoples Resources*)

Mack, E., H. Augare, L. Different Cloud-Jones, D. David, H. Quiver Gaddie, R. Honey, & R. Wippert. "Effective practices for creating transformative informal science education programs grounded in Native ways of knowing." *Cultural Studies of Science Education*, 7, 49-70. 2012.

Truth and Reconciliation Commission of Canada: Calls to Action. Truth and Reconciliation Commission of Canada, 2015. <http://www.trc.ca/websites/trcinstitution/File/2015/Findings/Calls_to_Action_English2.pdf>

Science First Peoples Teacher Resource Guide. First Nations Education Steering Committee and First Nations Schools Association, 2016.

Portage & Main Press, 2019 · Hands-On Science for British Columbia · Land, Water, and Sky for Grades K–2 · ISBN: 978-1-55379-797-5

How to Use *Hands-On Science* in Your Classroom

Hands-On Science is organized in a format that makes it easy for teachers to plan and implement. Four modules address the selected topics of study for kindergarten to grade-two classrooms. The modules relate directly to the Big Ideas, Core Competencies, Curricular Competencies, and Content outlined in the Science Curriculum for British Columbia.

Multi-Age Teaching and Learning

Whether working with students in a single-grade classroom from kindergarten to grade two, or working with multi-age classes, teachers will find appropriate learning opportunities in *Hands-On Science*. The lessons meet the diverse needs of all students through the implementation of differentiated instruction and personalized learning.

The Science Curriculum for British Columbia establishes specific Big Ideas, Curricular Competencies, and Content for each grade level. *Hands-On Science* has worked within themes to infuse these Big Ideas, Curricular Competencies, and Content into multi-age modules (see the Curriculum Learning Framework at the beginning of each module). It is therefore important for teachers to work collaboratively with their colleagues across grade levels to determine how best to implement lessons. The Curriculum Learning Frameworks will also be helpful, as each one includes a grade-level focus for specific lessons. This will assist teachers in both single-grade classrooms or multi-age classrooms to identify lessons and topics appropriate to their class.

Differentiated instruction and personalized learning will also ensure the needs of all students are met during science lessons. For example, in any classroom, whether multi-age or single-grade, students will be working at varying levels of literacy. As such, some students may be communicating their learning through drawing, while others may use single words, and yet others write several sentences. The lessons in *Hands-On Science* are developed to foster growth and learning at all literacy levels.

The same situation may be evident in terms of numeracy. For example, some students may be using comparative nonstandard measurement, while other students may be capable of working with standard metric measurement units and devices. There is plenty of flexibility in *Hands-On Science* to ensure that all students' learning needs can be met through active, student-centred learning.

Module Overview

Each module features an overarching question that fosters inquiry related to the Big Ideas. The module also has its own introduction, which summarizes the general concepts and goals for the module. This introduction provides background information for teachers, planning tips, and lists of vocabulary related to the module, as well as other pertinent information (e.g., how to embed Indigenous perspectives).

Also included at the beginning of each module is a Curriculum Learning Framework, which is based on the Big Ideas and Learning Standards (Curricular Competencies and Content) from the Science Curriculum for British Columbia (https://curriculum.gov.bc.ca/).

The Curriculum Learning Framework identifies the Big Ideas, Sample Guided Inquiry Questions, and Content for each grade level. As well, Content is connected to specific lessons, which are listed below each Content concept. Although specific lessons were intentionally written for grade-level content, much of this content is interconnected. As such, the overarching theme of the module provides a variety of connections to all three grade levels and, therefore, offers many springboards to learning.

▶

Portage & Main Press, 2019 · *Hands-On Science for British Columbia · Land, Water, and Sky for Grades K–2* · ISBN: 978-1-55379-797-5

Portage & Main Press, 2019 · Hands-On Science for British Columbia · Land, Water, and Sky for Grades K–2 · ISBN: 978-1-55379-797-5

Lesson Title
- provides a guided inquiry question related to the Learning Standards explored in the lesson

Information for Teachers
- presents basic scientific knowledge needed for activities

Explore
- presents whole-class and small-group activities which provide students with choice and opportunities to pose further inquiry questions while collaborating with peers
- details procedures, including higher-level questioning techniques, and suggestions for encouraging the development of concepts and skills
- identified as Explore Part One, Explore Part Two, and so on (when there is more than one in a lesson)

Expand
- provides opportunities for individual students to expand what they know, do, and understand
- empowers and encourages students to pose their own inquiry questions and conduct investigations, research, and projects individually, with support and facilitation by the teacher as needed; student success will depend on prior modelling, guided practice, and individual skills
- includes suggestions for Makerspace projects and Loose Parts exploration

Learning Centre
- supports diverse learners, promotes differentiated instruction, and is based on multiple-intelligences research (see page 17)
- includes a task card that remains at the centre, along with any required supplies and materials; review the task card before students work at the centre, to ensure they are familiar with the content and the expectations (students are not expected to read and comprehend all content on the card, but it serves as a guide for teachers and a visual prompt for students)

---

1 Initiating Event: What Do We Observe, Think, and Wonder About Plants and Animals?

Information for Teachers

In this lesson, students will participate in place-based learning to explore plants and animals in a local natural environment. Encourage students to suggest local natural areas, and plan ahead to select a location.

NOTE: It is important to prepare for guest speakers and to ensure that students are appropriately prepared as well. Review behavioural expectations and discuss questions that students may wish to ask the guest. Be sure to have students thank the speaker for the visit and consider following up with written or illustrated thank you notes. It is also important to consider protocols for Elders. Please see the *Science First Peoples Teacher Resource Guide* (see References, page xx) for guidelines and considerations.

In preparing to explore nature with students, consider referring to the book, *A Place for Wonder: Reading and Writing Nonfiction in the Primary Grades,* by Georgia Heard and Jennifer McDonough (see References, page xxx).

Materials
- chart paper
- markers
- digital cameras
- magnifying glasses
- tweezers
- stretch gloves
- recycled bags
- string
- portable whiteboard or chart paper with a sturdy backboard

Engage
Discuss the location for the place-based learning. Ask:
- Who has been to this place before?
- How did you get there?

- What is it like there?
- What do you think we will see there? Smell? Hear? Feel?

Introduce the guided inquiry question: **What do we observe, think, and wonder about plants and animals?**

Explore Part One
Once the class has arrived at the place-based learning location, provide time for students to explore the area freely (under adult supervision). Provide access to materials such as digital cameras, magnifying glasses, tweezers, stretch gloves, garden tools for exploration, and recycled bags in which to collect artifacts.

As students explore, pose questions for them to ponder. For example:
- What are you examining?
- Why is it interesting to you?
- What can you tell me about it?
- What do you wonder about it?
- What do you see? Feel? Smell? Hear?

Expand
To prepare students for journaling, do a journal entry together as a class. This will require a portable whiteboard and markers, or chart paper with a sturdy backboard.

- sketching or colouring plants and animals in the natural environment
- collecting flowers or leaves, sketching them, and then pressing them in the journal

Learning Centre
At the learning centre, provide a variety of objects and photographs collected during the place-based learning experience. Also provide magnifying glasses, tweezers, scissors, glue, art paper, poster paper, construction paper, art supplies, and a copy of the Learning-Centre

▶

2   **Hands-On Science: An Inquiry Approach** · Grades K–2

Materials
- lists all materials required to conduct the main activities
- includes items for display purposes or for recording students' ideas
- suggests visual materials (e.g., large pictures, sample charts, diagrams) to assist in presenting ideas and questions and encourage discussion
- connects to Image Bank visuals, which may be printed or projected for specific activities (see Appendix on page 165 for thumbnails and free access)

Engage
- activates prior knowledge, piques students' curiosity about related concepts, and introduces the lesson's guided inquiry question
- models for students how to pose their own inquiry questions; teachers may choose to record the guided inquiry question (e.g., on a sentence strip) for display, so students can refer to it during activities and discussions

Reproducibles

- may be used to guide activities or record data
- may also serve as a template for designing and constructing graphic organizers
- included as thumbnails in the lessons
- provided as full-sized, printable version on the Portage & Main website (see Appendix for URL and password)

Enhance

- enriches and elaborates on the Big Idea, Core Competencies, and Learning Standards with optional activities
- encourages active participation and learning through Family Connections

Embed Part One

- provides students with opportunities to participate in a Talking Circle (see page 16) to demonstrate their learning through consolidation and reflection
- allows for synthesis and application of inquiry and new ideas
- reviews main ideas of the lesson, focusing on the Big Idea, Core Competencies, and Learning Standards
- reviews guided inquiry question so students can share their knowledge, provide examples, and ask further inquiry questions

Embed Part Two

- embeds learning by adding to graphic organizers; having students record, describe, and illustrate new vocabulary; and adding new vocabulary to the word wall throughout the module or even all year
- provides opportunity to reflect the cultural diversity of the classroom and the community by including new terminology in languages other than English, including Indigenous languages
- explores Core Competencies with students to foster student self-assessment of how these skills were used throughout the lesson

Assessment

- provides suggestions for authentic assessment
- includes student self-assessment, formative assessment, and summative assessment (see pages 29–34)

1

How Can I Sort Objects From Nature?

1. Look at each object from our nature walk.
2. Describe how it looks, feels, smells, and sounds. (Do not taste it!)
3. Sort the objects into the bins.
4. Describe your sorting rules to others.

they demonstrated that competency. Ask questions directly related to that competency to inspire discussion. For example:

- Where did you get your ideas for your place-based journal entry today? (Creative Thinking)

Have students reflect orally, encouraging participation, questions, and the sharing of evidence (See page xx in the introduction for more information on these templates).

As part of this process, students can also set goals. Ask:

- What would you do differently next time and why?
- How will you know if you are successful in meeting your goal?

NOTE: Use the same prompts from these templates over time to see how thinking changes with different activities.

Embed Part One: Talking Circle

Revisit the guided inquiry question: **What do we observe, think, and wonder about plants and animals?** Have students share their experiences and knowledge, provide examples, and ask further inquiry questions.

Embed Part Two

- Focus on students' use of the Core Competencies. Have students reflect on how they used one of the Core Competencies (Thinking, Communicating, or Personal and Social skills) during the various lesson activities. Project one of the CORE COMPETENCY DISCUSSION PROMPTS templates (page xx-xx), and use it to inspire group reflection. Referring to the template, choose one or two "I Can" statements on which to focus. Then, have students use the "I Can" statements to provide evidence for how

Enhance

- **Family Connection:** Provide students with the following sentence starter:
  - A favourite place for us to visit outside is____.

  Have students take home the sentence starter to complete. Family members can help the student draw and write about this topic.

Student Self-Assessment (E)

Have students complete the COOPERATIVE SKILLS SELF-ASSESSMENT template, on page xx, to reflect on their success working with others, as they share and compare ideas.

Living Things   3

Portage & Main Press, 2019 · *Hands-On Science for British Columbia · Land, Water, and Sky for Grades K–2* · ISBN: 978-1-55379-797-5

The Curricular Competencies Correlation Chart at the beginning of each module provides details on how students' Curricular Competencies are developed through scientific inquiry. The chart outlines the skills, strategies, and processes that students use in the module and identifies the specific lessons in which these Curricular Competencies are the focus. The Curricular Competencies are developed in various ways over time, and therefore are addressed in multiple lessons throughout *Hands-On Science* modules.

Each module includes a list of related resources for students (books, websites, and online videos).

Each module is organized into lessons based on the Learning Standards. The first lesson in each module provides an initiating event, using an Observe-Think-Wonder strategy. Real-life explorations, often within the local environment, provide opportunity for place-based learning, which is discussed in more detail on page 18.

The second lesson in each module explores storytelling as it relates to the inquiry topics. This lesson includes an emphasis on Indigenous stories, children's literature, and nonfiction texts, while providing opportunities for students to engage in activities that focus on literacy and creative storytelling.

The last lesson in each module provides an opportunity for personalized learning through individualized inquiry, as students explore what more they would like to know, do, and understand about the module's Big Ideas.

Talking Circles

Talking Circles originated with First Nations leaders as a process to encourage dialogue, respect, and the co-construction of ideas. The following process is generally used in a Talking Circle:

- the group forms a complete circle
- one person holds an object such as a stick, feather, shell, or stone
- only the person holding the stick talks, while the rest listen
- the stick is passed around in a clockwise direction
- each person talks until they are finished, being respectful of time
- the Talking Circle is complete when everyone has had a chance to speak
- a person may pass the stick without speaking, if they choose

See <www.firstnationspedagogy.ca/circletalks.html> for more information. Also consider inviting a local Elder or Knowledge Keeper to share with the class the process of a Talking Circle.

Portage & Main Press, 2019 · Hands-On Science for British Columbia · Land, Water, and Sky for Grades K–2 · ISBN: 978-1-55379-797-5

Multiple Intelligences Learning Centres

Learning centres in **Hands-On Science** focus on a different multiple intelligence to provide opportunities for students to use areas of strength and also to expose them to new ways of learning.

Teachers are encouraged to explore the topic of multiple intelligences with their students and to have students self-reflect to identify ways they learn best, and ways that are challenging for them. Guidelines for this process are included in *Teaching to Diversity* by Jennifer Katz (see References, page 21).

Multiple Intelligence		These learners...
Verbal-Linguistic	V-L	...think in words and enjoy reading, writing, word puzzles, and oral storytelling.
Logical-Mathematical	L-M	...think by reasoning and enjoy problem solving, puzzles, and working with data.
Visual-Spatial	V-S	...think in visual pictures and enjoy drawing and creating visual designs.
Bodily-Kinesthetic	B-K	...think by using their physical bodies and enjoy movement, sports, dance, and hands-on activities.
Musical-Rhythmic	M-R	...think in melodies and rhythms and enjoy singing, listening to music, and creating music.
Interpersonal	INTER	...think by talking to others about their ideas and enjoy group work, planning social events, and taking a leadership role with friends or classmates.
Intrapersonal	INTRA	...think within themselves and enjoy quietly thinking, reflecting, and working individually.
Naturalistic	N	...learn by classifying objects and events and enjoy anything to do with nature and scientific exploration of natural phenomena.
Existential	EX	...learn by probing deep philosophical questions and enjoy examining the bigger picture as to why ideas are important.

▶

Portage & Main Press, 2019 · Hands-On Science for British Columbia · Land, Water, and Sky for Grades K–2 · ISBN: 978-1-55379-797-5

Icons

To provide a clear indication of important features of **Hands-On Science**, the following icons are used throughout lessons:

Place-Based Learning	Place-based learning focuses on the local environment and community. It is important for students to explore the local area in order to build personalized and contextual knowledge. Place-based learning: ■ emphasizes exploring the natural environment, replacing classroom walls with the natural land ■ offers firsthand opportunities to observe, explore, and investigate the land, waters, organisms, and atmosphere of the local region ■ promotes a healthy interplay between society and nature ■ helps students envision a world where there is meaningful appreciation and respect for our natural environment—an environment that sustains all life Many lessons in **Hands-On Science** incorporate place-based learning activities, whether it be a casual walk around the neighbourhood to examine trees or a more involved exploration of local waterways.
Applied Design, Skills, and Technologies	Throughout **Hands-On Science**, students have opportunities to use applied design, skills, and technologies to plan and construct objects. For example, in *Living Things for Grades K–2*, students design and construct models of an animal's environment to show how the animal meets its basic needs. Using applied design skills and technology, students seek solutions to practical problems through research and experimentation. There are specific steps: 1. Identify a need. Recognize practical problems and the need to solve them. 2. Create a plan. Seek alternate solutions to a given problem, create a plan based on a chosen solution, and record the plan through writing and labelled diagrams. 3. Develop a product or prototype. Construct an object that solves the given problem, and use predetermined criteria to test the product. 4. Communicate the results. Identify and make improvements to the product, and explain the changes.
Ecology and the Environment	**Hands-On Science** provides numerous opportunities for students to investigate issues related to ecology, the environment, and sustainable development. The meaning of sustainability can be clarified by asking students: "Is there enough for everyone, forever?" These topics also connect to Indigenous worldviews about respecting and caring for the Earth.
Technology	Digital learning, or information and communication technology (ICT), is an important component of any classroom. As such, technological supports available in schools—digital cameras, computers/tablets, interactive whiteboards (IWB), projectors, document cameras, audio-recording devices, calculators—can be used with and by students to enhance their learning experiences.
Classroom Safety	When there are safety concerns, teachers may decide to demonstrate an activity, while still encouraging as much student interaction as possible. The nature of science and scientific experimentation means that safety concerns do arise from time to time.

Portage & Main Press, 2019 · Hands-On Science for British Columbia · Land, Water, and Sky for Grades K–2 · ISBN: 978-1-55379-797-5

Makerspaces

To foster open inquiry and promote personalized learning, each module of **Hands-On Science** suggests a Makerspace as part of the Expand section. A Makerspace is a creative do-it-yourself environment, where participants pose questions, share ideas, and explore hands-on projects. In the school setting, a Makerspace is usually cross-curricular and should allow for inquiry, discovery, and innovation. Sometimes, the Makerspace is housed in a common area, such as the library, which means it is a space used by the whole school community. A classroom Makerspace is usually designed as a centre where students create do-it-yourself projects, emphasizing personalized learning, while collaborating with others on cross-curricular ideas. It is important to remember learning is not directed here. Rather, simply create conditions for learning to happen.

There is no list of required equipment that defines a Makerspace; however, the centre may evolve to foster inquiry within a specific topic. Students are given the opportunity to work with a variety of age-appropriate tools, as well as with everyday, arts-and-crafts, and recycled materials. Materials to consider at Makerspaces include:

- general supplies (e.g., graph or grid paper for planning and designing, pencils, markers, paper, cardstock, cardboard, scissors, masking tape, duct tape, glue, rulers, metre sticks, tape measures, elastic bands, string, Plasticine, modelling clay, fabric/cloth, straws, pipe cleaners, aluminum foil)
- recycled materials (e.g., various sizes of boxes, cardboard rolls, milk cartons, plastic bottles, spools, plastic lids)
- art supplies (e.g., paper, paint, markers, chalk, pastels, crayons, pencil crayons, beads, sequins, foam shapes, yarn, glass beads)
- building materials (e.g., sticks, wooden blocks, wooden dowels, toothpicks, craft sticks, balsa wood)
- age-appropriate tools (e.g., hammers, nails, screwdrivers, screws)
- natural objects (e.g., rocks, shells, feathers, seeds, wood slices, sticks)
- commercial products (e.g., LEGO, LEGO Story Starter, WeDo, MakeDo, Meccano, Plus-Plus, K'Nex, KEVA Planks, Dominoes, Wedgits)
- technology (e.g., Green Screen, iPads, coding/programming [Beebots, Code-a Pillar], apps such as Hopscotch, Tynker, Scratch Jr., Tickle)
- topic-based literature to inspire projects
- reference materials (e.g., books, videos, websites, visual images)

Work with students to develop a collaborative culture in which they tinker, invent, and improve on their creations. Ask students for ideas on how to stock the Makerspace, based on their project ideas, and then work collaboratively to acquire these supplies. The internet may also provide ideas for projects and materials.

Set up a recycling box/bin at the Makerspace for paper, cardboard, clean plastics, and other materials students can use for their creations. Stress to students that Makerspaces can help reuse many items destined for a landfill. Discuss which items can/should be placed in this bin.

Some things to consider when planning and developing a Makerspace are:

- Always address safety concerns, ensuring materials, equipment, and tools are safe for student use. Include safety gloves and goggles, as appropriate. Engage students in a discussion about safety and respect at the Makerspace before beginning each module. Consider sharp objects, small parts, and other potential hazards for students of

Portage & Main Press, 2019 · Hands-On Science for British Columbia · Land, Water, and Sky for Grades K–2 · ISBN: 978-1-55379-797-5

all ages and abilities who will have access to the Makerspace. At this age, this exploration needs to be supervised.

- Consider space and storage needs. Mobile carts and/or bins are handy for storing raw materials and tools.
- Work with students to write a letter to parents/guardians, explaining the purpose of the Makerspace, and asking for donations of materials.

In **Hands-On Science**, each module includes a variety of suggestions for Makerspace materials, equipment, possible challenges, and literature links related to the Big Ideas being explored.

The Makerspace process is intended for solving design problems, so it is helpful to have visuals at the Makerspace to encourage innovation, creativity, and the use of Applied Design, Skills, and Technologies (see page 18). In addition, although individual inquiry is encouraged, the Makerspace process is often collaborative in nature. Therefore, it is important to focus on skills related to working with others (see the Cooperative Skills Assessment templates on pages 49 and 51).

Before students begin working at a Makerspace, review Applied Design, Skills, and Technologies and collaborative skills with students. As a class, co-construct criteria for each skill, record on chart paper, and display at the Makerspace. Or, challenge students to create posters for the Makerspace that convey what Applied Design, Skills, and Technologies and collaboration look like. Refer to these visual prompts before, during, and after students work at the centre, as a means of guiding and assessing the process.

As students create, photograph their creations to share with the class, and discuss the unique properties of their designs. Model appropriate digital citizenship with students by asking their permission to photograph and share

their creations. Facilitate regular debriefing sessions as a class, after students have spent time at the Makerspace. Consider focusing this discussion on the Core Competencies (Thinking, Communication, and Personal and Social Skills) as an anchor for reflective practice.

The nature of a Makerspace is such that it provides an excellent venue for personalized learning. As students pose their own inquiry questions, they may choose to use the Makerspace to explore that question further.

Loose Parts

Closely related to the open inquiry fostered by the Makerspace, the theory of Loose Parts was first proposed back in the 1970s by architect Simon Nicholson. He believed it is the Loose Parts in our environment that empower our creativity. The theory has begun to influence early years educators intent on offering students opportunities to play freely with objects and materials, and to pose their own questions and investigations. Loose Parts include anything natural or synthetic (e.g., beads, buttons, fabric, washers and nuts, cardboard rolls, pom poms, acorns, leaves) that students can move, control, and manipulate. Loose Parts promote open-ended thinking that leads to problem solving, curiosity, and creativity. Play and learning possibilities are endless, as there is no single outcome that is achieved. Instead, Loose Parts offer opportunities for students to consider a wide range of possibilities and ideas.

When appropriate, provide provocations (questions to inspire play) that offer an entry point for a Loose Parts activity. As an example, while studying living things, teachers may provide bins of stones, twigs, bark, shells, and seed pods with the provocation, "How many different ways can you sort the objects?" Students may begin with such a sorting task, but expand to build structures, compare and

Portage & Main Press, 2019 · Hands-On Science for British Columbia · Land, Water, and Sky for Grades K–2 · ISBN: 978-1-55379-797-5

measure, or examine patterns on the various objects.

Throughout **Hands-On Science**, Loose Parts are used to engage students and as an opportunity to expand investigations, generate their own inquiry questions, and personalize learning. Suggestions for Loose Parts exploration are included in the Expand section of lessons. For more information about Loose Parts, see *Loose Parts: Inspiring Play in Young Children* by Lisa Daly and Miriam Beloglovsky and *Loose Parts: A Start-Up Guide* by Sally Haughey and Nicole Hill.

References

British Columbia Ministry of Education. *BC's New Curriculum.* 2016. <https://curriculum.gov.bc.ca/>

Daly, Lisa, and Miriam Beloglovsky. *Loose Parts: Inspiring Play in Young Children.* Redleaf Press, 2014.

Haughey, Sally, and Nicole Hill. *Loose Parts: A Start-Up Guide.* Fairy Dust Teaching, 2017.

Katz, Jennifer. *Teaching to Diversity.* Winnipeg, MB: Portage & Main Press, 2012.

"Talking Circles." *First Nations Pedagogy Online.* <www.firstnationspedagogy.ca/circletalks.html>

Portage & Main Press, 2019 · Hands-On Science for British Columbia · Land, Water, and Sky for Grades K–2 · ISBN: 978-1-55379-797-5

Curricular Competencies: How to Infuse Scientific Inquiry Skills and Processes Into Lessons

Hands-On Science is based on a scientific inquiry approach. While participating in the activities of **Hands-On Science**, students use a variety of scientific inquiry skills and processes as they answer questions, solve problems, and make decisions. These skills and processes are not unique to science, but they are integral to students' acquisition of scientific literacy. At the kindergarten to grade-two level, these include:

QP	questioning and predicting
PC	planning and conducting
PA	processing and analyzing data and information
AI	applying and innovating
C	communicating
E	evaluating

The icons above are used to link assessment suggestions to Curricular Competencies (see page 33 for more information).

Use the following guidelines to encourage the development of students' skills and processes in specific areas.

Observing

Students learn to perceive characteristics and changes through the use of all five senses. Encourage students to safely use sight, smell, touch, hearing, and taste to gain information about objects and events. Observations may be qualitative (e.g., texture, colour), quantitative (e.g., size, number), or both.

Observing includes:

- gaining information through the senses
- identifying similarities and differences, and making comparisons

Encourage students to communicate their observations in a variety of ways, including orally, in writing, by sketching labelled diagrams, and by capturing evidence digitally (e.g., with a digital camera).

Questioning

Generating thoughtful inquiry questions is an essential skill for students when participating in inquiry-based learning. Encourage students to be curious and to extend their questions beyond those posed to them.

Students should learn to formulate a specific question to investigate, one that can be answered through experimentation. This skill takes time to develop with young learners. Be patient, and provide the appropriate scaffolds as needed. Then students can create, from a variety of possible methods, a plan to find answers to the questions they pose.

Exploring

Students need ample opportunity to manipulate materials and equipment in order to discover and learn new ideas and concepts. During exploration, encourage students to use all of their senses and observation skills.

Oral discussion is an integral component of exploration; it allows students to communicate their discoveries. At a deeper level, discussion also allows students to make meaning by discussing inconsistencies and by comparing/contrasting their observations with others. This

Portage & Main Press, 2019 · Hands-On Science for British Columbia · Land, Water, and Sky for Grades K–2 · ISBN: 978-1-55379-797-5

is the constructivist model of learning, which is essential in inquiry-based learning. It is also essential to document the learning that is taking place for each child. This can be done through anecdotal observation records, photographs, videos, and interviews.

Classifying

Classification is used to group or sort objects and events, and is based on observable properties. For example, objects can be classified into groups according to colour, shape, or size. Two strategies for sorting include sorting mats and Venn diagrams. Sorting mats show distinct groups, while Venn diagrams intersect to show similar characteristics among sets.

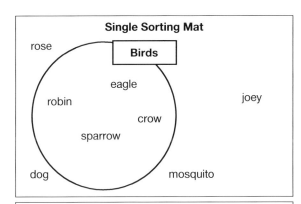

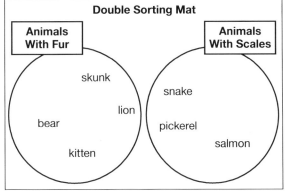

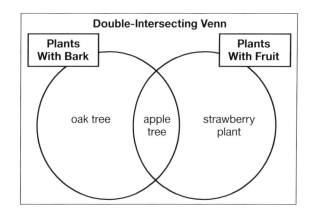

Measuring

Measuring is the process of discovering the dimensions or the quantity of objects or events and, at the kindergarten to grade-one level, usually involves comparing and ordering objects by length, area, volume, and mass. Measuring activities first involve the use of nonstandard units of measure (e.g., interlocking cubes, paper clips) to determine length. At the kindergarten to grade-one level, students generally use only nonstandard units to make simple measurements, as they build understanding of how to observe, compare, and communicate dimensions and quantity. This is a critical preface to measuring with standard units. By grade two, and/or in cases where students have demonstrated competence with nonstandard measurement, students may be introduced to standard units (e.g., centimetres, metres, grams, kilograms) and the use of measuring devices (e.g., metre sticks, tape measures, spring scales, calibrated beakers).

An essential skill of measurement is estimating. Encourage students to estimate before they measure, whether in nonstandard or standard units. Estimation gives students opportunities to take risks, use background knowledge, and enhance their measuring skills by comparing estimates and actual results.

▶

Portage & Main Press, 2019 · Hands-On Science for British Columbia · Land, Water, and Sky for Grades K–2 · ISBN: 978-1-55379-797-5

Communicating, Analyzing, and Interpreting

In science, communication is achieved through diagrams, graphs, charts, maps, models, and symbols, as well as with written and spoken languages. At the kindergarten to grade-two level, communicating includes:

- viewing images
- making labelled diagrams*
- journaling
- reading and interpreting data from simple tables and charts
- making tables and charts
- reading and interpreting data from pictographs
- making pictographs
- making models
- using oral languages
- sequencing and grouping events, objects, and data according to attributes

*NOTE: Depending on students' literacy skills, they may label diagrams with letters and words. Model this to show that a scientific diagram includes accurate illustrations, labels, and sometimes measurements.

Journaling is an important strategy that offers students an opportunity to communicate their ideas, understandings, and questions through emergent writing and illustrating. A journal can be used to summarize activities, share new knowledge, or record observations.

A journal is especially useful during place-based learning experiences. Scientists, such as Charles Darwin, a naturalist, and Rachel Carson, a marine biologist and conservationist, used this strategy for recording observations. The **place-based journal** encourages students to slow down, pay attention to the world around them, heighten their senses, observe more accurately, and reflect on experiences (Bell, 2017). See About This Module, page 60, for more information.

Other forms of communication in science include reading and interpreting charts and graphs. When presenting students with charts and graphs, or when students make their own as part of a specific activity, there are guidelines that should be followed:

- A **pictograph** has a title and information on one axis that denotes the items being compared (note that the first letter of each word on both the title and the axis text is capitalized). There is generally no graduated scale or heading for the axis representing numerical values.

Favourite Dessert				
		🍦		
		🍦		
	🥧	🍦		
🍰	🥧	🍦		
Cake	Pie	Ice Cream		

- A **tally chart** is a means of recording data as an organized count. The count is grouped in fives for ease of determining the total by counting by fives.

Favourite Sport		
Sport	**Tally**	**Total**
baseball	ⵘⵘ	6
hockey	ⵘⵘ ⵘⵘ	10
soccer	ⵘⵘ ⵘⵘ ||	12

Portage & Main Press, 2019 · Hands-On Science for British Columbia · Land, Water, and Sky for Grades K–2 · ISBN: 978-1-55379-797-5

- A **chart (table)** requires an appropriate title, and both columns and rows need specific headings. Again, all titles and headings require capitalization of the first letter of each word, as in the title of a story. In some cases, pictures can be used to make the chart easier for young students to understand. Charts can be made in the form of checklists or can include room for additional written information and data.

Checklist Chart

Which Substances Dissolve in Water?		
Substance	Dissolves in Water	Does Not Dissolve in Water
Beads		√
Sugar	√	
Drink Mix	√	
Rice		√
Pepper		√

Data Chart

Local Snowfall		
Month	2016/2017 Snowfall (cm)	Average Snowfall (cm)
October	7	5
November	9	8
December	23	20
January	29	25
February	16	18
March	11	10

Communicating also involves using the language and terminology of science. Encourage students to use the appropriate vocabulary related to their investigations (e.g., *object, metal, heavy, strong, movement*). The language of science also includes terms such as *predict, infer, estimate, measure, experiment*, and *hypothesize*. Use this vocabulary regularly throughout all activities and encourage students to do the same. In each module, work with students to develop a word wall on which to record terms students have learned. Have students provide visuals, and define the terms in their own words.

Predicting

Predicting refers to the question, "What do you think will happen?" For example, ask students to predict what they think will happen to an inflated balloon that is placed in a basin of water. It is important to provide opportunities for students to make predictions and to feel safe doing so.

Inferring

In a scientific context, inferring refers to deducing why something occurs. For example, ask students to infer why an inflated balloon floats when placed into a basin of water. Again, it is important to encourage students to take risks when making inferences. Instead of explaining scientific phenomena to them, give students opportunities to infer for themselves, using a variety of perspectives, and then build their knowledge base through inquiry and investigation.

Inquiry Through Investigation and Experimentation

When investigations and experiments are conducted in the classroom, planning and recording both the process and the results are essential. The traditional scientific method uses the following format:

- **purpose:** what we want to find out, or a testable question we want to answer
- **hypothesis:** a prediction; what we think will happen, and why
- **materials:** what we used to conduct the experiment or investigation
- **method:** what we did
- **results:** what we observed and measured
- **conclusion:** what we found out
- **application:** how we can use what we learned

Portage & Main Press, 2019 · Hands-On Science for British Columbia · Land, Water, and Sky for Grades K–2 · ISBN: 978-1-55379-797-5

This method of recording investigations may be used in later school years. However, in the early grades, it may be more appropriate to focus on a narrative style of reporting such as:

- what we want to know
- what we think might happen
- what we used
- what we did
- what we observed
- what we found out

A simpler four-question narrative may also be used with any age group. The structure includes the following questions:

- What was I looking for? (Describe the question you were trying to answer, or the hypothesis/prediction you were testing.)
- How did I look for it? (Tell what you did. Include materials and method.)
- What did I find? (Describe observations and data.)
- What does this mean? (Draw conclusions, and consider applications to real life.)

This narrative may be done in a variety of ways: oral discussion as a class, recording findings as a class, having students use drawings, or a combination of these.

Throughout **Hands-On Science**, a variety of methods are used to encourage students to communicate the inquiry process, including those above. Other formats such as concept maps and other graphic organizers are also used.

Inquiry Through Research

In addition to hands-on inquiry, research is another aspect of inquiry that involves finding, organizing, and presenting information related to a specific topic or question. Scientific inquiry involves making observations, exploring, asking questions, and looking for answers to those questions. Even at a young age, students can begin to research topics studied in class if they are provided with support and guidelines.

Accordingly, guided research is a teaching and learning strategy encouraged throughout **Hands-On Science**. Guided research provides an opportunity for students to seek further information about subjects of inquiry, personal interests, or topics of their choice. As such, students are empowered and engaged in the process.

Guided research encourages students and teachers to do the following:

Students	Teachers
■ ask questions of interest related to a topic being studied by the class ■ choose resources ■ collect information ■ make a plan to present findings ■ present research in a variety of ways	■ provide opportunities for students to ask questions of personal interest ■ provide access to resources ■ model and support the research process ■ offer opportunities for students to present their findings in a variety of ways and to a variety of audiences

In **Hands-On Science**, the approach to inquiry and research is one of gradual release. In order to provide more opportunity for success and independence in conducting research, it is important to scaffold the process. Consider using gradual release of responsibility (Regie Routman, 2008) as a learning model for research:

Portage & Main Press, 2019 · Hands-On Science for British Columbia · Land, Water, and Sky for Grades K–2 · ISBN: 978-1-55379-797-5

I Do It	▪ Begin by modeling the process. Select an inquiry question and demonstrate how to choose resources. ▪ Use the resources available in the class science library. Also use selected websites appropriate for your students (see Resources for Students at the beginning of each module). ▪ Next, demonstrate how to collect information through note taking (jot notes), labelled diagrams, pictures, or photographs. ▪ Finally, make a plan to present and display findings for the class.
We Do It	▪ Next, have the class choose another inquiry question. ▪ Together, choose resources and collect information. ▪ Together, make a plan to present findings and display findings for another class or for family/community members.
You Do It	▪ Students now choose their own inquiry questions, conduct research, and present findings.

Throughout this gradual release process, the teacher provides substantial support in initial inquiry experiences and, over time, presents students with more and more opportunities for directing their own research.

NOTE: Given that this module of *Hands-On Science* was developed for students in kindergarten to grade two, research needs to be age-appropriate. Collect a variety of resources at early reading levels, including visuals, books, and websites. Students can then access information and record their ideas using pictures, labelled diagrams, and simple texts, such as jot notes. If, for example, a student is researching how an eagle builds a nest, they may examine a variety of pictures, books, websites, and videos, and then create a series of labelled drawings to present their findings.

As students build research skills and progress through the grades, research will increase in sophistication. The guided approach and the

I Do, We Do, You Do model will support students with independent inquiry as they continue to do scientific research in higher grades. For more information about this model (also known as the Optimal Learning Model), see *Teaching Essentials: Expecting the Most and Getting the Best from Every Learner, K-8* by Regie Routman, "Gradual Release of Responsibility: I do, We do, You do" by E. Levy, and "Teaching New Concepts: 'I Do It, We Do It, You Do It' Method" by A. McCoy.

Addressing Students' Early Literacy Needs

The inquiry process involves having students ask questions and conduct investigations and research to answer these questions. At the kindergarten to grade-two level, students will benefit from literacy support while conducting research. Consider having volunteers, student mentors, or educational assistants provide support during these processes to help young students conduct appropriate research and communicate their findings orally or visually. Also consider conducting brief lessons on how to read and glean information from pictures when investigating questions through the inquiry process. As well, students can show their learning through labelled pictures, invented/temporary spelling, or the use of technology that allows them to record their research orally.

Online Considerations

As our technological world continues to expand at an accelerating rate, and information is increasingly available online, students will turn to the internet more and more to expand their learning. Accordingly, **Hands-On Science** is replete with opportunities for students to use online resources for research and investigation. Discuss online safety protocols with students. Be vigilant in supervising student use of the internet. Similarly, review websites and bookmark those appropriate for student use.

Portage & Main Press, 2019 · *Hands-On Science for British Columbia · Land, Water, and Sky for Grades K–2* · ISBN: 978-1-55379-797-5

Also discuss plagiarism with students: copying information word for word—whether from a book, the internet, or another resource—is wrong. Such information should always be paraphrased in the student's own words, and the source of the information cited. Photographs, drawings, figures, and other images found online should also only be used with permission and citation of the source. Alternatively, students can source images for which permission has already been granted, such as through Creative Commons. Creative Commons is a non-profit organization that "promotes and enables the sharing of knowledge and creativity…[and which] produces and maintains a free suite of licensing tools to allow anyone to easily share, reuse, and remix materials with a fair 'some rights reserved' approach to copyright." See <creativecommons.org>.

References

Banchi, Heather, and Randi Bell. "The Many Levels of Inquiry," *Science and Children* 46.2 (2008): 26–29.

Bell, Antonella. *Nature Journaling*. University of Alberta, 2017.

British Columbia Ministry of Education. *BC's New Curriculum*. 2016. <https://curriculum.gov.bc.ca/>

Creative Commons. <creativecommons.org>

Davies, Anne. *Making Classroom Assessment Work* (3rd ed.). Courtenay, BC: Connections Publishing, 2011.

First Nations Schools Association. *Science First Peoples Teacher Resource Guide* (2016).

Fullan, Michael. *Great to Excellent: Launching the Next Stage of Ontario's Education Agenda* (2013).

Levy, E. (2007). Gradual Release of Responsibility: I do, We do, You do. <www.sjboces.org/doc/Gifted/GradualReleaseResponsibilityJan08.pdf>

Katz, Jennifer. *Teaching to Diversity: The Three-Block Model of Universal Design for Learning.* Winnipeg: Portage & Main Press, 2012.

Manitoba Education and Training. *Kindergarten to Grade 4 Science: Manitoba Curriculum Framework of Outcomes*, 1999. (See: <www.edu.gov.mb.ca/>)

McCoy, Antoine (2011, March 4). Teaching New Concepts: "I Do It, We Do It, You Do It" Method. <antoinemccoy.com/teaching-new-concepts>

Ontario Literacy and Numeracy Secretariat. "Inquiry-based Learning," Capacity Building series 32, p. 4 (May 2013).

Routman, Regie. *Teaching Essentials: Expecting the Most and Getting the Best from Every Learner, K–8*. Portsmouth, NH: Heinemann, 2008.

Toulouse, Pamela. *Achieving Aboriginal Student Success*. Winnipeg: Portage & Main Press, 2011.

Truth and Reconciliation Commission of Canada: *Calls to Action*, 2015. <www.trc.ca>

Portage & Main Press, 2019 · Hands-On Science for British Columbia · Land, Water, and Sky for Grades K–2 · ISBN: 978-1-55379-797-5

The *Hands-On Science* Assessment Plan

Hands-On Science provides a variety of assessment tools that enable teachers to build a comprehensive and authentic daily assessment plan for students. Based on current research about the value of quality classroom assessment (Davies, 2011), suggestions are provided for authentic assessment, which includes student self-assessment and reporting of Core Competencies.

British Columbia's K–12 Assessment System (see <https://curriculum.gov.bc.ca/assessment-system> and <https://curriculum.gov.bc.ca/classroom-assessment-and-reporting>) states:

> Assessment and curriculum are interconnected. Curriculum sets the learning standards that give focus to classroom instruction and assessment. Assessment involves the wide variety of methods or tools that educators use to identify student learning needs, measure competency acquisition, and evaluate students' progress toward meeting provincial learning standards.
>
> [British Columbia's] assessment system is being redesigned to align with the new curriculum. Assessment of all forms will support a more flexible, personalized approach to learning and measure deeper, complex thinking. [British Columbia's] educational assessment system strives to support student learning by providing timely, meaningful information on student learning through multiple forms of assessment. The assessment system has three programs:
>
> 1. Classroom Assessment and Reporting
> 2. Provincial Assessment
> 3. National and International Assessment
>
> Classroom assessment is an integral part of the instructional process and can serve as a meaningful source of information about student learning. Feedback from ongoing assessment in the classroom can be immediate and personal for a learner and guide the learner to understand their [strengths and challenges] and use the information to set new learning goals.

The primary purpose of assessment is to improve student learning. *Hands-On Science* provides assessment suggestions, rubrics, and templates for use during the teaching/learning process. These assessment suggestions include tasks related to *student self-assessment* of the Core Competencies, as well as *formative assessment* and *summative assessment* by the teacher.

Student self-assessment helps students develop their capacity to set their own goals, monitor their own progress, determine their next steps in learning, and reflect on their learning in relation to the three Core Competencies—Thinking, Communication, and Social and Personal.

Formative assessment requires that teachers provide students with descriptive feedback and coaching for improvement in relation to the Learning Standards (Curricular Competencies and Content).

Summative Assessment is comprehensive in nature, and is intended to identify student progress in relation to the Learning Standards (Curricular Competencies and Content).

Both summative and formative assessments are an integral part of a balanced classroom assessment plan. Then, when student self-assessment is infused in this assessment plan, a clearer picture emerges of where a student is in relation to the Core Competencies and Learning Standards.

Student Self-Assessment

It is important for students to reflect on their own learning. For this purpose, a variety of assessment templates are provided in *Hands-On Science*. Depending on their literacy levels, students may complete self-assessments in various ways. For example, the templates may be used as guides for oral conferences between teacher and student, or an adult may act as a

Portage & Main Press, 2019 · *Hands-On Science for British Columbia · Land, Water, and Sky for Grades K–2* · ISBN: 978-1-55379-797-5

scribe for the student, recording their responses. As well, students can show their learning through labelled pictures, invented/temporary spelling, and writing, with guidance and support as needed.

For the purpose of self-assessment, find a STUDENT SELF-ASSESSMENT template, on page 35, as well as a STUDENT REFLECTIONS template on page 36.

The SCIENCE JOURNAL, on page 37, will encourage students to reflect on their own learning. Print several copies for each student, cut the pages in half, add a cover, and bind the pages together. Students can then create their own title pages for their journals. For variety, have students use the blank reverse side of each page for other reflections, such as drawing or writing about:

- new challenges
- favourite activities
- real-life experiences
- new terminology
- new places explored during investigations

Students may also journal in other ways, such as by adding notes to their portfolios, or by keeping online science blogs or journals to record successes, challenges, and next steps related to learning goals.

NOTE: This SCIENCE JOURNAL template is provided as a suggestion, but journals can also be made from simple notebooks or recycled paper.

Another component of student self-assessment involves opportunities for students to reflect on their use of the Core Competencies. During each lesson, spend time discussing and reflecting on one of the Core Competencies. The intent here is to enhance students' ability to recognize how and when they use the competencies during the inquiry process. Reflection on Core Competencies is ongoing, since students' strengths and challenges in using the Core

Competencies may differ in various contexts and activities.

For the purpose of this assessment process, project a copy of one of the five CORE COMPETENCY DISCUSSION PROMPTS templates on page 38–42 (one each on Communication, Creative Thinking, Critical Thinking, Positive Personal and Cultural Identity, and Personal Awareness and Responsibility). Choose one or two "I Can" statements on which to focus discussion. Students then use the "I Can" statements to provide evidence of how they demonstrated that competency (model this process for the class). For example, a student might say:

- I can ask and answer questions. I know this because I asked lots of questions about the Sun. I also answered questions about how to stay safe by wearing sunscreen, sunglasses, and hats.

The intent is to provide an opportunity for group discussion and modelling, while encouraging individual students to reflect on their use of the Core Competencies. Choose the Core Competencies and facets that are most appropriate for each lesson. There is lots of room for differentiating based on the strengths and needs of the class.

NOTE: Although the facets are identified on these templates, they are featured only for teacher reference. Students are not expected to refer to the facets during reflective discussion.

To inspire students to further reflect on each Core Competency, use a variety of self-reflection prompts. For this purpose, use the CORE COMPETENCY SELF-REFLECTION FRAMES on pages 43–47 throughout the learning process. There are five frames provided to address the Core Competencies, one each on Communication, Creative Thinking, Critical Thinking, Positive Personal and Cultural Identity, and Personal Awareness and Responsibility. Conference

Portage & Main Press, 2019 · Hands-On Science for British Columbia · Land, Water, and Sky for Grades K–2 · ISBN: 978-1-55379-797-5

individually with students to support self-reflection, or students may complete prompts using words and pictures.

NOTE: Use the same prompts from these templates over time, to see how thinking changes with different activities.

Another component of student self-assessment utilizes the CORE COMPETENCY STUDENT REFLECTIONS: MODULE SUMMARY template, on page 48. This is completed by students at the end of a module, in order to encourage them to reflect on how their Core Competencies have developed over time. Students' reflections are recorded in the rectangle on the template. Then, the student considers next steps in learning as related to that particular Core Competency. These reflections are recorded on the arrow on the template, again, using words and drawings.

NOTE: It is important to keep in mind that the Core Competencies will only be self-assessed by students, and not directly assessed by teachers. However, teachers may conference with students in order to encourage them to think about and discuss their learning over time.

Students should also be encouraged to reflect on their cooperative group work skills, since these are directly related to Core Competencies, as well as to the skills scientists use as they collaborate in team settings. For this purpose, a COOPERATIVE SKILLS SELF-ASSESSMENT template is on page 49.

Student reflections can also be done in many other ways. For example, students can:

- interview one another
- write an outline or script and make a video
- create a slide show with an audio recording

Formative Assessment

It is important to assess students' understanding before, during, and after a lesson. The information gathered helps determine students' needs in order to plan the next steps in instruction. Students may come into class with misconceptions about science concepts. By identifying what they already know, teachers can help students make connections and address any challenges.

Formative assessment provides opportunities for teachers to document evidence of each student's learning. Along with utilizing evidence gathered from photographs, videos, and digital portfolios, document evidence of learning by using the formative assessment templates provided in *Hands-On Science*.

To assess students as they work, use the formative assessment suggestions provided with many of the activities. While observing and conversing with students, use the ANECDOTAL RECORD template and/or the INDIVIDUAL STUDENT OBSERVATIONS template to record assessment data.

- **Anecdotal Record**: To gain an authentic view of a student's progress, it is critical to record observations during lessons. The ANECDOTAL RECORD template, on page 50, provides a format for recording individual or group observations.
- **Individual Student Observations**: To focus on individual students for a longer period of time, consider using the INDIVIDUAL STUDENT OBSERVATIONS template, on page 51. This template provides more space for comments and is especially useful during conferences, interviews, or individual student performance tasks. It is important to note that not every student has to be observed during the same lesson. Observations can take place over time in order to focus on each student's learning.

Formative assessment also involves the consideration of students' collaborative skills. Always assess a student's individual

Portage & Main Press, 2019 · *Hands-On Science for British Columbia · Land, Water, and Sky for Grades K–2* · ISBN: 978-1-55379-797-5

performance, not the work of a group. Assess how an individual student works within a group. Such skill development includes the ability to use words and actions to encourage other students, contribute to group work, and use strengths and skills to complete a given task (British Columbia Ministry of Education, 2016). For this purpose, use the COOPERATIVE SKILLS TEACHER ASSESSMENT template on page 52. Use this template as a checklist or for anecdotal comments.

Both formative assessment and summative assessment include *performance assessment*. Performance assessment is planned, systematic observation and assessment based on students actually doing a specific science activity. A SAMPLE RUBRIC and a RUBRIC template for teacher use are on pages 54 and 53. For any specific activity, before the work begins, discuss and co-construct with students success criteria for completing the task. This will ensure the success criteria relate to the lesson's learning goals. Record these criteria on the rubric. Use the rubric criteria to assess student performance, using the proficiency scale from the British Columbia Ministry of Education Framework for Classroom Assessment (see References, page 34):

Emerging	The student demonstrates an initial understanding of the concepts and competencies relevant to the expected learning.
Developing	The student demonstrates a partial understanding of the concepts and competencies relevant to the expected learning.
Proficient	The student demonstrates a complete understanding of the concepts and competencies relevant to the expected learning.
Extending	The student demonstrates a sophisticated understanding of the concepts and competencies relevant to the expected learning.

Observe student during the performance task being assessed to determine their level of proficiency on each criterion (see SAMPLE RUBRIC, page 54). Share this data with students to provide descriptive feedback, and to encourage student reflection related to performance task criteria and the level of proficiency.

Summative Assessment

Summative assessment provides a summary of student progress related to the Learning Standards at a particular point in time. It is important to gather a variety of assessment data to draw conclusions about what a student knows, can do, and understands. As such, consider collecting student products, observing processes, and having conversations with students. Only the most recent and consistent evidence should be used.

Summative assessment suggestions are provided with the culminating lesson of each module of **Hands-On Science**. Use the ANECDOTAL RECORD template, found on page 50, the INDIVIDUAL STUDENT OBSERVATIONS template, found on page 51, and the RUBRIC, found on page 53, to record student results.

A student portfolio is another format that can be used for summative assessment. A portfolio is a collection of work that shows evidence of a student's learning. There are many types of portfolios—the showcase portfolio and the progress portfolio are two popular formats. *Showcase portfolios* highlight the best of students' work, with students involved in the selection of pieces and justification for choices. *Progress portfolios* reflect students' progress as their work improves and aim to demonstrate in-depth understanding of the materials over time. Select, with student input, work to include in a science portfolio or in a science section of a multi-subject portfolio. Selections should include

Portage & Main Press, 2019 · *Hands-On Science for British Columbia · Land, Water, and Sky for Grades K–2* · ISBN: 978-1-55379-797-5

representative samples of student work in all types of science activities.

Templates are included to organize the portfolio (PORTFOLIO TABLE OF CONTENTS, page 55, and PORTFOLIO ENTRY RECORD, page 56).

Indigenous Perspectives on Assessment

From an Indigenous perspective, assessment is community-based, qualitative, and holistic, and includes input from all the people who influence an individual student's learning— parents, caregivers, Elders, Knowledge Keepers, community members, and educators. An assessment that includes all these perspectives provides a balanced understanding of what represents success for Indigenous students and their families/community. A strong partnership between parents/guardians/communities and school improves student achievement. Be aware that some Indigenous students may feel apprehensive about a formal process of assessment; others may find that Western achievement goals do not fit their worldview.

In **Hands-On Science**, consideration has been given to assessment from an Indigenous perspective. The following suggestions will assist in supporting this perspective:

- Consider learning and assessment in a holistic way, acknowledging that each student will find identity, meaning, and purpose through connections to the community, to the natural world, and to values such as respect and gratitude.

- Incorporate family and community in learning and assessment. Include parents/caregivers, siblings, grandparents, aunts and uncles, and cousins. Also include community members, such as Elders, Knowledge Keepers, daycare staff, babysitters, and coaches. For this purpose, a template is included for FAMILY

AND COMMUNITY CONNECTIONS: ASSESSING TOGETHER, which is found on page 57. After any lesson or module, students can take home a copy of this template to complete with family or community members (with permission). This template can also be completed by students in pairs, to enhance the sense of community in the classroom.

- Have students take home one of their self-assessment templates (STUDENT SELF-ASSESSMENT, STUDENT REFLECTIONS, SCIENCE JOURNAL, CORE COMPETENCY SELF-REFLECTION FRAMES, CORE COMPETENCY STUDENT REFLECTIONS: MODULE SUMMARY, or COOPERATIVE SKILLS SELF-ASSESSMENT) to explain it to a family or community member. These templates can also be shared with a peer to enhance the sense of community within the school.

Connecting Assessment to Curricular Competencies

Throughout **Hands-On Science**, suggestions are provided for student self-assessment, formative assessment, and summative assessment. Many of these suggestions are linked to the Curricular Competencies, as in the following example that focuses on Communication:

Formative Assessment

- Photograph students as they journal, to collect evidence of learning activities. Be sure to document student thinking after journaling. For example, meet with them individually to have them share their thoughts about the journaling experience and what they recorded in their journal. Use photographs taken as they journal to inspire reflection. Focus on students' ability to express and reflect on personal experiences of place. Use the INDIVIDUAL STUDENT OBSERVATIONS template on page 51 to record interview highlights.

Portage & Main Press, 2019 · *Hands-On Science for British Columbia · Land, Water, and Sky for Grades K–2* · ISBN: 978-1-55379-797-5

This feature of the **Hands-On Science** Assessment Plan supports teachers in making connections between assessment strategies and the Curricular Connections focused upon at the kindergarten to grade-two levels.

Module Assessment Summary

At the end of each module, suggestions are provided for a summary of assessment. This includes:

- Collecting student work in a portfolio, so students can examine and discuss these artifacts of learning during a conference.
- Having students take home a copy of the FAMILY AND COMMUNITY CONNECTIONS: ASSESSING TOGETHER template on page 57 to complete with a family or community member.
- Having students complete the CORE COMPETENCY STUDENT REFLECTIONS: MODULE SUMMARY template, on page 48, to reflect on their use of the Core Competencies throughout the module and to determine next steps in their learning.
- Reviewing assessment templates completed by students and teachers throughout the module.

Important Note to Teachers

It is important to keep in mind that the ideas provided in **Hands-On Science** for student self-assessment, formative assessment, and summative assessment are merely suggestions. Teachers are encouraged to use the assessment strategies presented in a wide variety of ways, and to ensure that they build an effective assessment plan using these assessment ideas, as well as their own valuable experiences as educators.

References

British Columbia Ministry of Education. *A Framework for Classroom Assessment.* <https://curriculum.gov.bc.ca/sites/curriculum.gov.bc.ca/files/pdf/assessment/a-framework-for-classroom-assessment.pdf>

British Columbia Ministry of Education. *B.C. Performance Standards.* <https://www2.gov.bc.ca/gov/content/education-training/k-12/teach/bc-performance-standards>

British Columbia Ministry of Education. *BC's New Curriculum.* 2016. <https://curriculum.gov.bc.ca/>

British Columbia Ministry of Education. *Supporting the Self-Assessment and Reporting of Core Competencies*, 2016 <https://curriculum.gov.bc.ca/sites/curriculum.gov.bc.ca/files/pdf/supporting-self-assessment.pdf>.

Cameron, Caren, and Kathleen Gregory. *Rethinking Letter Grades: A Five-Step Approach for Aligning Letter Grades to Learning Standards.* Winnipeg: Portage & Main Press, 2014.

Davies, Anne. *Making Classroom Assessment Work* (3rd ed.). Courtenay, BC: Connections Publishing, 2011.

Manitoba Education. *Rethinking Classroom Assessment with Purpose in Mind: Assessment for Learning, Assessment as Learning, Assessment of Learning*, 2006.

Ontario Ministry of Education. *Growing Success: Assessment, Evaluation, and Reporting in Ontario Schools*, 2010. <www.edu.gov.on.ca/>.

Toulouse, Pamela. *Achieving Aboriginal Student Success.* Winnipeg: Portage & Main Press, 2011.

Portage & Main Press, 2019 · Hands-On Science for British Columbia · Land, Water, and Sky for Grades K–2 · ISBN: 978-1-55379-797-5

Date: _____   Name: _____

Student Self-Assessment

Looking at My Science Learning

1. Today in science, I _____

_____

_____

2. In science, I learned _____

_____

_____

3. I did very well at _____

_____

_____

4. One science skill that I am working on is _____

_____

_____

5. I would like to learn more about _____

_____

_____

6. One thing I like about science is _____

_____

_____

Note: The student may complete this self-assessment or the teacher can scribe for the student.

Portage & Main Press, 2019 · *Hands-On Science for British Columbia · Land, Water, and Sky for Grades K–2* · ISBN: 978-1-55379-797-5

Date: _____   **Name:** _____

Student Reflections

What I Did	What I Learned

Next Steps in My Learning	My Strengths and Challenges

Portage & Main Press, 2019 · Hands-On Science for British Columbia · Land, Water, and Sky for Grades K–2 · ISBN: 978-1-55379-797-5

Science Journal

Date: _____   Name: _____

[box]

Today, I _____

I learned _____

I would like to learn more about _____

Science Journal

Date: _____   Name: _____

[box]

Today, I _____

I learned _____

I would like to learn more about _____

Portage & Main Press, 2019 · Hands-On Science for British Columbia · Land, Water, and Sky for Grades K–2 · ISBN: 978-1-55379-797-5

Core Competency Discussion Prompts

Communication

Facet	I Can...
A. **Connect and engage with others (to share and develop ideas)**	■ ask questions ■ answer questions ■ be an active listener (My eyes are on the speaker. I show that I am interested in what they are saying.) ■ see that my classmates and I can sometimes have different ways to do things, see things, and understand things ■ use a calm voice when I disagree with others
B. **Acquire, interpret, and present information (includes inquiries)**	■ understand and share information about a topic that is important to me ■ present information clearly and in an organized way ■ present information and ideas to an audience I may not know
C. **Collaborate to plan, carry out, and review constructions and activities**	■ work with others; I do my share of my group's job ■ take on roles and responsibilities in a group ■ describe ideas ■ explain the ways my group agrees with our ideas
D. **Explain/recount and reflect on experiences and accomplishments**	■ give feedback to my classmates ■ listen to feedback from my classmates ■ use feedback from my classmates to make changes to my ideas ■ share simple experiences and activities and tell something I learned ■ show my learning ■ tell how my learning is connected to my experiences and hard work

Core Competency Discussion Prompts
Thinking: Critical Thinking

Facet	I Can...
A. **Analyze and critique**	■ show if I like something or not ■ use criteria ■ look at results from different points of view ■ reflect on and evaluate my thinking, products, and actions ■ think about my own beliefs and consider views that do not fit with them
B. **Question and investigate**	■ explore materials and actions ■ ask questions that have more than one answer ■ gather information ■ carefully think about different ways to solve a problem ■ decide if my sources of information are dependable ■ tell the difference between facts and opinions
C. **Develop and design**	■ experiment with different ways of doing things ■ help create criteria for design projects ■ keep track of my progress ■ change my actions to make sure I reach my goal ■ make choices that will help me meet my goals

Portage & Main Press, 2019 · Hands-On Science for British Columbia · Land, Water, and Sky for Grades K–2 · ISBN: 978-1-55379-797-5

Core Competency Discussion Prompts
Thinking: Creative Thinking

Facet	I Can...
A. **Novelty and value**	■ get ideas when I play (My ideas are fun for me and make me happy.) ■ get new ideas or build on other people's ideas, to create new things ■ think of new ideas as I follow my interests ■ think of ideas that are new to my classmates ■ make creative projects in an area that interests me
B. **Generating ideas**	■ get ideas when I use my senses to explore ■ build on others' ideas ■ add new ideas of my own to create new things ■ add new ideas of my own to solve problems ■ learn a lot about something (e.g., by doing research, talking to others, practising), so I can create new ideas ■ calm my mind (e.g., walking away for a while, doing something relaxing, being playful), so I can be more creative ■ have interests that I continue for a long time
C. **Developing ideas**	■ make my ideas work or I change what I am doing ■ usually make my ideas work with materials if I keep playing with them ■ learn and use the skills I need to make my ideas work, and usually succeed, even if it takes a few tries ■ use my experiences for future learning ■ try to develop my ideas over a long period of time

Core Competency Discussion Prompts
Personal and Social: Positive Personal & Cultural Identity

Facet	I Can...
A. **Relationships and cultural contexts**	■ describe my family ■ describe my community ■ tell about the different groups that I belong to ■ understand that my identity is made up of life experiences, family history, heritage, and peer groups ■ understand that learning is forever, and I will continue to grow as a person
B. **Personal values and choices**	■ tell what is important to me ■ explain my values and how they affect choices I make ■ tell how some important parts of my life have influenced my values ■ understand how my values affect my choices
C. **Personal strengths and abilities**	■ describe my characteristics ■ describe my talents and the things I do well ■ think about the things I do well ■ describe how I am a leader in my community ■ understand that I will continue to develop new strengths and skills to help me meet new challenges

Portage & Main Press, 2019 · Hands-On Science for British Columbia · Land, Water, and Sky for Grades K–2 · ISBN: 978-1-55379-797-5

Core Competency Discussion Prompts
Personal and Social: Personal Awareness and Responsibility

Facet	I Can...
A. **Self-determination**	■ be happy and proud of how well I did ■ celebrate my hard work and success ■ believe in myself ■ believe in my ideas ■ imagine and work toward change in myself and the world ■ learn about things in which people have different opinions
B. **Self-regulation**	■ sometimes name different emotions ■ use strategies that help me manage my feelings and emotions ■ work through hard tasks ■ make a plan and evaluate the results ■ take responsibility for my goals ■ take responsibility for my own learning ■ take responsibility for my own behaviour
C. **Well-being**	■ participate in activities that are healthy for my mind and body ■ tell/show how these healthy activities help me ■ take some responsibility for caring for my own body and mind ■ make choices that are good for my mind and body and keep me safe in my community, including my online conversations with others ■ use strategies to find peace when I am feeling stress ■ live a healthy life that includes both work and play

Date: _____   Name: _____

Core Competency Self-Reflection Frame
Communication

I Can...	Examples	Next Steps
I can answer questions.		
I can listen to others when they speak.		
I can share my learning.		
I can work in a group.		

Portage & Main Press, 2019 · *Hands-On Science for British Columbia · Land, Water, and Sky for Grades K–2* · ISBN: 978-1-55379-797-5

Portage & Main Press, 2019 · Hands-On Science for British Columbia · Land, Water, and Sky for Grades K–2 · ISBN: 978-1-55379-797-5

Date: _____

Name: _____

Core Competency Self-Reflection Frame
Creative Thinking

I can get new
ideas as I learn.

I can make
my ideas work.

I can learn
a lot through play.

I can learn
new skills as I try out my ideas.

One thing that I would like to work on is _____

Name: _____

Date: _____

Core Competency Self-Reflection Frame
Critical Thinking

I can explore materials.

I can experiment with different ways of doing things.

I can show if I like something or not.

I can ask questions and gather information.

One goal that I have is

Portage & Main Press, 2019 · Hands-On Science for British Columbia · Land, Water, and Sky for Grades K–2 · ISBN: 978-1-55379-797-5

Portage & Main Press, 2019 · Hands-On Science for British Columbia · Land, Water, and Sky for Grades K–2 · ISBN: 978-1-55379-797-5

Date: _____

Name: _____

Core Competency Self-Reflection Frame
Positive Personal and Cultural Identity

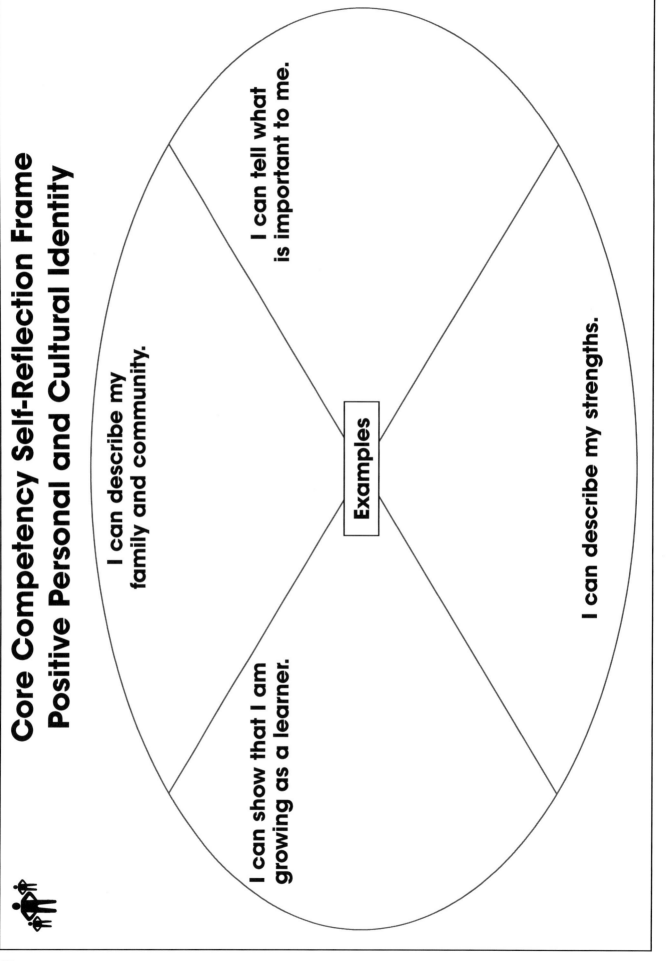

I can describe my family and community.

I can tell what is important to me.

I can describe my strengths.

I can show that I am growing as a learner.

Examples

Date: _____   **Name:** _____

Core Competency Self-Reflection Frame
Personal Awareness and Responsibility

I can be proud when I have done well.

I can describe my feelings.

I can work hard to finish a job.

I can make choices that make me feel good and stay safe.

Portage & Main Press, 2019 · *Hands-On Science for British Columbia · Land, Water, and Sky for Grades K–2* · ISBN: 978-1-55379-797-5

Portage & Main Press, 2019 · Hands-On Science for British Columbia · Land, Water, and Sky for Grades K–2 · ISBN: 978-1-55379-797-5

Date: _____

Name: _____

Core Competency Student Reflections

Module Summary

Core Competency: _____

What I Did

Next Steps

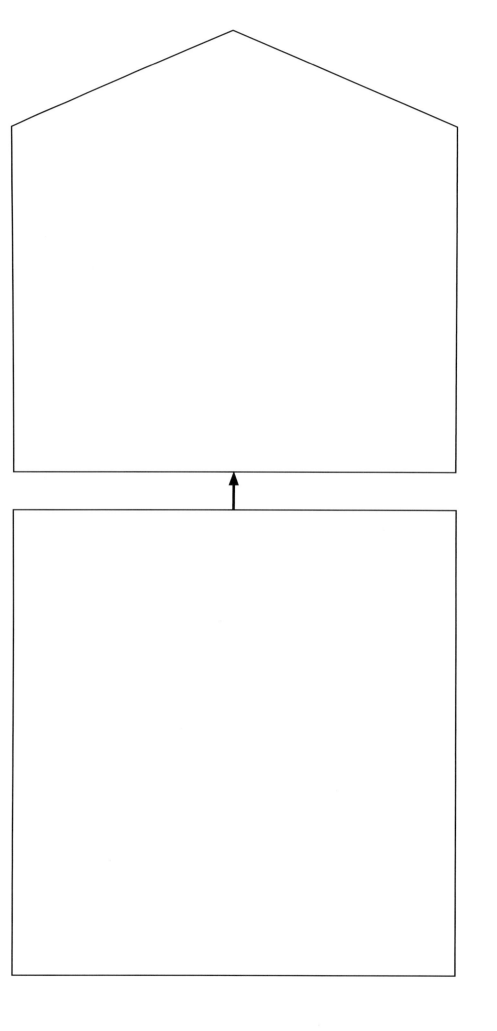

Date: _____   Name: _____

Cooperative Skills Self-Assessment

Students in my group:

_____   _____

_____   _____

Group Work—How Did I Do Today?

Group Work	How I Did (✔)		
I shared ideas.			
I listened to others.			
I asked questions.			
I encouraged others.			
I helped with the work.			
I stayed on task.			

I did very well at _____

_____

Next time, I would like to do better at _____

_____

Portage & Main Press, 2019 · Hands-On Science for British Columbia · Land, Water, and Sky for Grades K–2 · ISBN: 978-1-55379-797-5

Anecdotal Record

Purpose of Observation: _____

Student/Group	Date	Student/Group	Date
Comments		**Comments**	
Student/Group	**Date**	**Student/Group**	**Date**
Comments		**Comments**	
Student/Group	**Date**	**Student/Group**	**Date**
Comments		**Comments**	

Portage & Main Press, 2019 · Hands-On Science for British Columbia · Land, Water, and Sky for Grades K–2 · ISBN: 978-1-55379-797-5

Individual Student Observations

Purpose of Observation: _____

Student:	**Date:**
Observations:	

Student:	**Date:**
Observations:	

Student:	**Date:**
Observations:	

Portage & Main Press, 2019 · Hands-On Science for British Columbia · Land, Water, and Sky for Grades K–2 · ISBN: 978-1-55379-797-5

Portage & Main Press, 2019 · *Hands-On Science for British Columbia · Land, Water, and Sky for Grades K–2* · ISBN: 978-1-55379-797-5

Cooperative Skills Teacher Assessment

Date: _____

Task: _____

Group Member	Cooperative Skills				
	Contributes ideas and questions	Respects and accepts contributions of others	Negotiates roles and responsibilities of each group member	Remains focused and encourages others to stay on task	Completes individual commitment to the group

Rubric

Activity: _____

Module: _____

Date: _____

E – Emerging
D – Developing
P – Proficient
EX – Extending

Student	Criteria						

Portage & Main Press, 2019 · Hands-On Science for British Columbia · Land, Water, and Sky for Grades K–2 · ISBN: 978-1-55379-797-5

Portage & Main Press, 2019 · Hands-On Science for British Columbia · Land, Water, and Sky for Grades K–2 · ISBN: 978-1-55379-797-5

Sample Rubric

Activity: Looking at Seeds

Module: Living Things

Date: _____

E – Emerging
D – Developing
P – Proficient
EX – Extending

Criteria

Student	Observes Seeds Carefully	Asks Questions About the Seeds	Sorts Seeds and Gives Sorting Rules	Describes Seeds in Detail
Jarod	P	P	D	P
Aisha	P	D	P	D

SAMPLE

Name: _____

Portfolio Table of Contents

Entry	Date	Selection
1.	_____	_____
2.	_____	_____
3.	_____	_____
4.	_____	_____
5.	_____	_____
6.	_____	_____
7.	_____	_____
8.	_____	_____
9.	_____	_____
10.	_____	_____
11.	_____	_____
12.	_____	_____
13.	_____	_____
14.	_____	_____
15.	_____	_____
16.	_____	_____
17.	_____	_____
18.	_____	_____
19.	_____	_____
20.	_____	_____

Portage & Main Press, 2019 · Hands-On Science for British Columbia · Land, Water, and Sky for Grades K–2 · ISBN: 978-1-55379-797-5

Date: _____   **Name:** _____

Portfolio Entry Record

This work was chosen by _____

This work is _____

_____

I chose this work because _____

_____

_____

_____

_____

Note: The student may complete this form or the teacher can scribe for the student.

✂ -

Date: _____   **Name:** _____

Portfolio Entry Record

This work was chosen by _____

This work is _____

_____

I chose this work because _____

_____

_____

_____

_____

Note: The student may complete this form or the teacher can scribe for the student.

Portage & Main Press, 2019 · Hands-On Science for British Columbia · Land, Water, and Sky for Grades K–2 · ISBN: 978-1-55379-797-5

Family and Community Connections: Assessing Together

Family/Community Member's Name: _____

Draw a picture that shows what you have been learning in science. Work together to label your picture and describe your learning in words.

```



_____

_____

_____
```

What do you like best about what you have been learning in science?

_____

_____

What does your family/community member like best about what you have been learning in science?

_____

_____

_____

Portage & Main Press, 2019 · Hands-On Science for British Columbia · Land, Water, and Sky for Grades K–2 · ISBN: 978-1-55379-797-5

What Are the Features of the Land, Water, and Sky?

About This Module

This module of **Hands-On Science** focuses on characteristics of the land, water, and sky. Students will conduct investigations that explore the following Big Ideas:

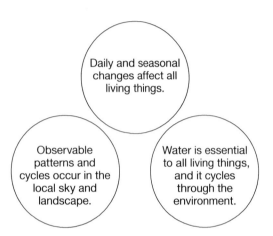

Daily and seasonal changes affect all living things.

Observable patterns and cycles occur in the local sky and landscape.

Water is essential to all living things, and it cycles through the environment.

While investigating these Big Ideas, the Curricular Competencies will be addressed as students use the following skills, strategies, and processes:

QP	questioning and predicting
PC	planning and conducting investigations
PA	processing and analyzing data and information
AI	applying and innovating
C	communicating
E	evaluating

In this module of **Hands-On Science**, students focus on changes in cycles, as they pertain to daily and seasonal patterns, as well as patterns in the sky, landscape, and water. Students will actively investigate these changes over time. For this module, it may be appropriate to focus on the suggested activities throughout the year, as opposed to teaching the module in one shorter block of time. This is especially true when studying weather, the months, and seasons.

Incorporate Indigenous perspectives and worldviews into lessons whenever possible.

■ Ensure Indigenous worldviews are represented in classroom learning (e.g., when exploring concepts related to daily and seasonal changes).

■ Learn how humans, animals, and plants depend on the Sun for light and for warmth to meet their basic needs for survival.

■ Discuss with students how having a respectful relationship with the environment is essential to understanding the seasonal changes that impact our daily lives.

■ Develop a better understanding of weather conditions and how to identify changes of the seasons.

■ Understand that Indigenous peoples often have an intimate relationship with all nature, including the Sun. For example, they understand the importance of the Sun in determining such things as the time of day or how much daylight is available to complete daily tasks. They also understand that available daylight changes with the four seasons. Sundials placed in strategic locations will tell time; placement varies, depending on the location of the community (longitude/latitude) in relation to the Sun.

When implementing place-based learning, consider Indigenous perspectives and knowledge. Outdoor learning provides an excellent opportunity to identify the importance of place. For example, use a map of the local area to have students identify the location of the placed-based learning in relation to the school.

Portage & Main Press, 2019 · Hands-On Science for British Columbia · Land, Water, and Sky for Grades K–2 · ISBN: 978-1-55379-797-5

This will help students develop a stronger image of their community and surrounding area.

Identify on whose traditional territory the school is located, the traditional territory of the location for the place-based learning (if different), as well as the traditional names for both locations. The following map, "First Nations in British Columbia," from Indigenous Services Canada can be used for this purpose: <www.aadnc-aandc.gc.ca/DAM/ DAM-INTER-BC/STAGING/texte-text/ inacmp_1100100021016_eng.pdf>

Incorporate land acknowledgment when students have learned on whose territory the school and place-based learning location are located. The following example can be used for guidance:

■ We would like to acknowledge that we are gathered today on the traditional, ancestral, and unceded territory of the _____ people, in the place traditionally known as _____.

NOTE: Many school districts have established protocols for land acknowledgment. Check with colleagues who support Indigenous education to see if there are specific protocols to follow.

Planning Tips for Teachers

■ Collect a variety of relevant reading materials, at a range of reading levels, to allow students to engage with the materials more frequently and to help them build their knowledge base. If possible, suggest websites students can visit, and allow students to peruse these on their own time. Always preview any websites students may use.

■ Review the Resources for Students list on page 67, and access any available resources, as well as any alternate resources on the same topics.

■ Read through the learning-centre activities in each lesson ahead of time, and prepare

any necessary materials, supplies, and resources.

■ Have students make place-based journals for use throughout the module. These can be made from notebooks with sturdy covers or from drawing paper and clip boards. Use a zipper-lock plastic bag to carry journal supplies (e.g., pencils, sharpeners, pencil crayons [rather than markers which will bleed if wet], stretch gloves for journaling on cooler days).

■ Record each lesson's guided inquiry question (e.g., on a sentence strip) for display throughout related investigations.

Loose Parts

Create Loose Parts bins to explore objects and materials. Fill bins with various objects and materials such as:

■ feathers
■ seeds and pine cones
■ sticks
■ soil samples
■ seasonal objects (e.g., mittens, toques, sun glasses, flip-flops)
■ shells
■ rocks
■ glass and plastic beads
■ plastic animals
■ crayons

Loose Parts bins encourage personalized learning, as students can select bins to explore. Offer provocations to inspire students (e.g., How can you use Loose Parts to show how land, water, and sky are important in your life?). Loose Parts can also expand inquiry; for example, students might explore a bin of shells and then choose to initiate an inquiry to find out the origin of various shells in the collection.

For more information about Loose Parts, see page 20.

Portage & Main Press, 2019 · *Hands-On Science for British Columbia · Land, Water, and Sky for Grades K–2* · ISBN: 978-1-55379-797-5

Makerspace

Develop a Makerspace. Classroom Makerspaces are usually designed as centres where students learn together and collaborate on do-it-yourself projects. Students are given the opportunity to work with a variety of age-appropriate tools, as well as everyday and recycled materials. Additionally, arts-and-crafts are often integrated into Makerspace offerings.

For this module, set up a Makerspace in your classroom that encourages informal learning about land, water, and sky. Include general materials, such as those listed in the Introduction on page 19, as well as module-specific materials, including measuring cups, funnels, siphons, magnifying glasses, and different shaped containers for holding water.

Do-it-yourself projects may include anything related to the concepts within this module. Projects students might initiate include (but are not limited to):

- designing and building a model of a house that uses sunlight for light and heat
- designing and building a birdhouse for a winter bird
- designing and building a sundial
- designing and building models or a diorama of trees in the different seasons
- creating a personalized calendar
- creating a device that helps to reduce water usage around the home (e.g., new type of sprinkler)
- creating a device or product that helps clean the water in a community
- creating a simple water filter that turns murky/dirty water into cleaner-looking water, using items from home (e.g., coffee filters, plastic bowls, cups, straws, sand)
- creating a model river ecosystem, using clay, plastic wrap, sand, and reusable plastic containers

- creating an ad campaign or puppet show to bring awareness to water conservation or water pollution
- creating a plant container that conserves water

Literacy connections that might inspire projects include:

- *The Rabbit Problem* by Emily Gravett
- *When the Wind Stops* by Charlotte Zolotow and Stefano Vitale
- *Sadie and the Snowman* by Allen Morgan
- *Painting the Wind* by Patricia and Emily MacLachlan
- *The Wind Blew* by Pat Hutchins
- *The Boy Who Harnessed the Wind* by William Kamkwamba

As inquiry questions are posed with each lesson, you will find these questions inspire other do-it-yourself projects related to the module. Students may determine solutions to these questions through the creating they do at the Makerspace. Remember to not direct the learning here; simply create conditions for learning to happen.

For more information about Makerspaces, see page 19.

Science Vocabulary

Throughout this module, use, and encourage students to use, vocabulary such as:

- *afternoon, behaviour, characteristic, cloud, condensation, condense, cycle, day, daytime, dew, dormant, energy, evaporate, evaporation, evening, fall, fog, freeze, frost, hail, heat, hibernate, lake, light, living thing, marsh, melt, months of the year, moon, morning, night, nighttime, ocean, planet, pollution, pond, puddle, rain, rainbow, river, sea, season, sleet, snow, source(s) of drinking water, spring, star, stream, summer,*

Portage & Main Press, 2019 · Hands-On Science for British Columbia · Land, Water, and Sky for Grades K–2 · ISBN: 978-1-55379-797-5

Sun, survival, temperature, water cycle, water usage, waterfall, weather, wind, winter

Infuse vocabulary related to scientific inquiry skills into daily lessons. This vocabulary could be displayed on the classroom throughout the year, as it relates to all science Big Ideas. Have students brainstorm which skills they are being asked to use as they work in particular lessons. They can also discuss how the skill looks and sounds as they explore and investigate. Vocabulary related to scientific inquiry skills include:

- *access, ask, brainstorm, collect, compare, connect, consider, construct, cooperate, create, describe, estimate, explain, explore, find, follow, graph, identify, improve, investigate, match, measure, observe, order, plan, predict, recognize, record, repeat, research, respond, select, sequence, test*

Early in the module, create a word wall for the module. The word wall can be created on a bulletin board or simply on poster or chart paper. Record new vocabulary on the bulletin board or poster as it is introduced during the module. Ensure the word wall is placed in a location in the classroom where all students can see and refer to the words during activities and discussion. Have students work with the terms on a regular basis by creating their own definitions, giving examples, linking terms in sentences, and using terms in context.

NOTE: Include terminology in languages other than English on the word wall. This is a way of acknowledging and respecting students' cultural backgrounds, while enhancing learning for all students.

A variety of online dictionaries may be used as a source for translations. For example:
- <http://www.freelang.net/online/haida.php>
- <http://www.firstvoices.com/en/Halqemeylem>

Online dictionaries are also available for languages other than English that may be reflective of the class cultural makeup.

Portage & Main Press, 2019 · *Hands-On Science for British Columbia · Land, Water, and Sky for Grades K–2* · ISBN: 978-1-55379-797-5

Curriculum Learning Framework

	K	1	2
Big Idea	Daily and seasonal changes affect all living things.	Observable patterns and cycles occur in the local sky and landscape.	Water is essential to all living things, and it cycles through the environment.
Possible Guiding Inquiry Questions	▪ What daily and seasonal changes can you see or feel? ▪ How are plants and animals affected by daily and seasonal changes?	▪ What kinds of patterns in the sky and landscape are you aware of? ▪ How do patterns and cycles in the sky and landscape affect living things?	▪ Why is water important for all living things? ▪ How can you conserve water in your home and school? ▪ How does water cycle through the environment?

Content	▪ weather changes [lesson 1, 2, 3, 6, 7, 19]
	▪ seasonal changes [lesson 6, 8, 9, 10, 11]
	▪ living things make changes to accommodate daily and seasonal cycles [lesson 4, 5, 8, 9, 10, 11]
	▪ First Peoples knowledge of seasonal changes [lesson 6, 8, 9, 10, 11]
		▪ common objects in the sky [lesson 4, 6, 12, 13, 19]
		▪ knowledge of First Peoples
		  ▪ shared First Peoples' knowledge of the sky [lesson 6, 12, 13]
		  ▪ local First Peoples' knowledge of the local landscape, plants and animals [lesson 8, 9, 10, 11, 19]
		  ▪ local First Peoples' understanding and use of seasonal rounds [lesson 11]
		▪ local patterns that occur on Earth and in the sky [lesson 4, 5, 8, 9, 10, 11, 12, 13]
			▪ water sources including local watersheds [lesson 14, 15, 16, 19]
			▪ water conservation [lesson 16, 17, 18]
			▪ the water cycle [lesson 15, 19]
			▪ local First Peoples' knowledge of water:
			  ▪ water cycles
			  ▪ conservation
			  ▪ connection to other systems [lesson 14, 15, 16, 17, 18, 19]

Core Competencies	*Thinking* *Communicating* *Social and Personal*		

Portage & Main Press, 2019 · *Hands-On Science for British Columbia · Land, Water, and Sky for Grades K–2* · ISBN: 978-1-55379-797-5

Hands-On Science for British Columbia

Curricular Competencies Correlation Chart

Throughout this module, students will develop Curricular Competencies by participating in learning experiences that focus on specific skills, strategies, and processes. The chart below represents the multiple opportunities students have to explore the Curricular Competencies.

Curricular Competencies	Lesson																		
	1	2	3	4	5	6	7	8	9	10	11	12	13	14	15	16	17	18	19
(QP) Questioning and Predicting																			
Demonstrate curiosity and a sense of wonder about land, water, and sky.	√	√	√	√	√	√	√	√	√	√	√	√	√	√	√	√	√	√	√
Observe land, water, and sky in familiar contexts.	√	√	√	√	√	√	√	√	√	√	√	√	√	√	√	√	√	√	√
Ask simple questions about land, water, and sky.	√	√	√	√	√	√	√	√	√	√	√	√	√	√	√	√	√	√	√
Make simple predictions about land, water, and sky.	√	√		√		√	√		√		√						√	√	
(PC) Planning and conducting investigations																			
Make exploratory observations using their senses.	√	√	√	√	√	√	√	√	√	√	√	√	√	√	√	√	√	√	√
Record observations.	√		√	√	√	√	√	√	√		√	√	√	√	√	√	√	√	√
Safely manipulate materials.	√	√	√	√	√	√	√	√	√	√	√	√	√	√	√	√	√	√	√
Make simple measurements using nonstandard units.						√	√												
(PA) Processing and analyzing data and information																			
Experience and interpret the local environment.	√							√	√			√		√					
Recognize First Peoples stories (including oral and written narratives), songs, and art, as ways to share knowledge.	√	√	√	√	√	√	√	√	√	√	√	√	√	√	√	√	√	√	√
Discuss observations about land, water, and sky.	√	√	√	√	√	√	√	√	√	√	√	√	√	√	√	√	√	√	√
Represent observations and ideas by drawing charts and simple pictographs.	√			√		√	√			√	√		√	√				√	√

▶

Portage & Main Press, 2019 · Hands-On Science for British Columbia · Land, Water, and Sky for Grades K–2 · ISBN: 978-1-55379-797-5

Curricular Competencies	Lesson																		
	1	2	3	4	5	6	7	8	9	10	11	12	13	14	15	16	17	18	19
Sort and classify data and information using drawings, pictographs and provided tables.				√	√		√		√	√	√								
Compare observations with predictions through discussion.	√					√	√					√					√	√	
Identify simple patterns and connections related to land, water, and sky.				√	√		√	√	√	√	√	√		√					

(AI) Applying and innovating

	1	2	3	4	5	6	7	8	9	10	11	12	13	14	15	16	17	18	19
Take part in caring for self, family, classroom, and school through, personal approaches.							√				√					√		√	
Transfer and apply learning to new situations.		√	√	√	√	√	√	√	√	√	√	√	√	√	√	√	√	√	√
Generate and introduce new or refined ideas when problem-solving.	√	√	√	√	√	√	√	√	√	√	√	√	√	√	√	√	√	√	√

(C) Communicating

	1	2	3	4	5	6	7	8	9	10	11	12	13	14	15	16	17	18	19
Share observations and ideas using oral written language, drawing, or role-play.	√	√	√	√	√	√	√	√	√	√	√	√	√	√	√	√	√	√	√
Express and reflect on personal experiences of place.	√						√	√				√		√					

(E) Evaluating

	1	2	3	4	5	6	7	8	9	10	11	12	13	14	15	16	17	18	19
Compare observations of land, water, and sky with those of others.	√	√		√			√	√	√	√		√	√		√	√	√	√	√
Consider some environmental consequences of their actions as related to land, water, and sky.	√						√	√				√		√			√	√	

Portage & Main Press, 2019 · Hands-On Science for British Columbia · Land, Water, and Sky for Grades K–2 · ISBN: 978-1-55379-797-5

Resources for Students

NOTE: Resources marked with an asterisk are considered to be authentic resources, meaning they reference the Indigenous community they came from, they state the individual that shared the story and gave permission for the story to be used publicly, and the person who originally shared the story is Indigenous. Stories that are works of fiction were written by an Indigenous author. For more information, please see *Authentic First Peoples Resources* at: <http://www.fnesc.ca/learningfirstpeoples/>

Books

*Aleck, Celestine. *Strong Stories Coast Salish: Why Ravens and Wolves Hunt Together.* Nanaimo, BC: Strong Nations Publishing, 2016.

*——.*The Sun and the Moon.* Nanaimo, BC: Strong Nations Publishing, 2016.

*Armstrong, Jeanette. *Neekna and Chemai.* Penticton, BC: Theytus Books, 2008.

Ball, Nate. *Let's Investigate with Nate #1: The Water Cycle.* New York: HarperCollins, 2017.

Birnbaum, Abe. *Green Eyes.* Victoria, BC: Dragonfly Books, 2011.

*Boreham, Brenda and Terri Mack. *From the Mountains to the Sea: We Are a Community.* Nanaimo, BC: Strong Nations Publishing, 2015

Boreham, Brenda. *Strong Readers Northern Series: Look Up at the Sky!* Nanaimo, BC: Strong Nations, 2014.

*Bouchard, David. *Night and Day.* Turtle Voices series. Newmarket, ON: Pearson, 2011.

Bouchard, David. *Rainbow Crow.* Markham, ON: Red Deer Press, 2012.

Branley, Franklyn M. *Down Comes the Rain.* New York: HarperCollins, 2017.

Branley, Franklyn M. *The Big Dipper.* New York: HarperCollins, 1991.

Brown, Margaret Wise. *Goodnight Moon.* New York: HarperCollins, 2007.

Brummel, Connie. *Maple Moon.* Markham, ON: Fitzhenry & Whiteside, 1999.

Bushey, Jeanne. *Orphans in the Sky.* Markham, ON: Red Deer Press, 2004.

*Buck, Wilfred. *Tipiskawi Kisik: Night Sky Star Stories.* Winnipeg, MB. Manitoba First Nations Education Resource Centre, 2018.

Carle, Eric. *Little Cloud.* New York: Penguin Random House, 1996.

Carle, Eric. *Papa, Please Get the Moon for Me.* New York: Simon & Schuster, 1986.

*Calve, Genevieve, Albert Dumont, and Joan Commanda Tenasco. *The Maple Leaves of Kitchi Makwa.* Ottawa: Turtle Moons Contemplations, 2011.

Cherry, Lynne. *A River Ran Wild: An Environmental History.* New York: HMH Books, 2002.

Children of La Loche and Friends (eds.). *Byron Through the Seasons: A Dene-English Story Book.* Saskatoon, SK: Fifth House Publishers, 1994.

Christopher, Neil, David Natcher, and Mary Ellen Thomas. *Arctic Weather and Climate Through the Eyes of Nunavut's Children.* Iqaluit, NU: Inhabit Media, 2013.

Clarke, George Elliott. *Lasso the Wind: Aurelia's Verses and other Poems.* Nanaimo, BC: Strong Nations Publishing, 2013.

Cole, Joanna. *The Magic School Bus at the Waterworks.* New York: Scholastic, 1986.

Portage & Main Press, 2019 · *Hands-On Science for British Columbia · Land, Water, and Sky for Grades K–2* · ISBN: 978-1-55379-797-5

*Corneau, Michelle. *Creation Story: Sky Woman*. Nanaimo, BC: Strong Nations Publishing, 2016.

*——. *Strong Stories Kanyen'keha:ka: Big Bear*. Nanaimo, BC: Strong Nations Publishing, 2016.

Cox Cannons, Helen. *Clouds*. Mankato, MN: Capstone Publishing, 2015.

——. *Wind*. Mankato, MN: Capstone Publishing, 2015.

Dean, Janice. *Freddy the Frogcaster and the Big Blizzard*. Washington, DC: Regnery Publishing, 2014.

——. *Freddy the Frogcaster and the Huge Hurricane*. Washington, DC: Regnery Publishing, 2015.

——. *Freddy the Frogcaster and the Terrible Tornado*. Washington, DC: Regnery Publishing, 2016.

——. *Freddy the Frogcaster*. Washington, DC: Regnery Publishing, 2013.

de Coteau, Sandra Orie. *Did You Hear the Wind Sing Your Name?: An Oneida Song of Spring*. New York: Bloomsbury Press, 2002.

dePaola, Tomie. *Michael Bird-Boy*. Englewood Cliffs, NJ: Prentice Hall, 1975.

Ducharme, Linda. *The Bannock Book*. Winnipeg, MB: Pemmican Publications, 2009.

Flaherty, Louise. *Things That Keep Us Warm*. Iqaluit, NU: Inhabit Education, 2016.

*Flett, Julie. *Owls See Clearly at Night*. Vancouver, BC: Simply Read Books, 2010.

*——. *Wild Berries*. Vancouver, BC: Simply Read Books, 2013.

Gibbons, Gail. *It's Raining*. New York: Holiday House Publishing, 2015.

*Gillespie, Desiree. *A Journey Through the Circle of Life*. Winnipeg, MB: Pemmican Publications, 2015.

Graham, Bob. *Home in the Rain*. Somerville, MA: Candlewick Press, 2017.

Gravett, Emily. *The Rabbit Problem*. New York: Simon & Schuster, 2010.

*Guiboche, Audrey. *Kawlija's Blueberry Promise*. Winnipeg, MB: Pemmican Publications, 2015.

Gürth, Per-Henrik. *Canada All Year*. Toronto: Kids Can Press, 2011.

*Hainnu, Rebecca. *A Walk on the Shoreline*. Toronto: Inhabit Education, 2015.

Hanna, Julie. *The Man Who Named the Clouds*. Park Ridge, IL: Albert Whitman, 2006.

Heard, Georgia and Jennifer McDonough. *A Place for Wonder*. Portsmouth, NH: Stenhouse Publishers, 2009.

Henkes, Kevin. *Old Bear*. New York: Greenwillow Books, 2008.

Hicks, Nola. *Hurry Up, Ilua!*. Iqaluit, NU: Inhabit Media, 2015.

Hutchins, Pat. *The Wind Blew*. Toronto: Simon & Schuster Canada, 1993.

Ikuutaq, Paula Rumbolt. *The Legend of Lightning and Thunder*. Iqaluit, NU: Inhabit Media, 2013.

Jeffers, Oliver. *How to Catch a Star*. New York: HarperCollins, 2015.

Jeffers, Oliver. *The Way Back Home*. New York: Philomel Books, 2008.

Joe, Donna. *Salmon Boy: A Legend of the Sechelt People*. Gibsons, BC: Nightwood Editions, 2001.

Portage & Main Press, 2019 · Hands-On Science for British Columbia · Land, Water, and Sky for Grades K–2 · ISBN: 978-1-55379-797-5

Kamkwamba, William. *The Boy Who Harnessed the Wind*. Toronto: HarperCollins Canada, 2010.

Kooser, Ted. *Bag in the Wind*. Somerville, MA: Candlewick Press, 2010.

MacLachlan, Patricia and Emily MacLachlan. *Painting the Wind*. Toronto: HarperCollins Canada, 2006.

Maclear, Kyo. *The Fog*. Toronto: Tundra Publishing, 2017.

McCartney, Tania. A *Canadian Year: Twelve Months in the Life of Canadian Kids*. Toronto: HarperCollins, 2017.

*McLeod, Elaine. *Lessons From Mother Earth*. Toronto: House of Anansi Press, 2010.

Meeks, Arone Raymond. *Enora and the Black Crane*. Broome, AU: Magabala Books, 2010.

Mike, Nadia. *Seasonal Cycles*. Iqaluit, NU: Inhabit Education, 2016.

Miller, Debbie S. *Arctic Lights, Arctic Nights*. New York: Bloomsbury Press, 2007.

Miyakoshi, Aikko. *The Storm*. Toronto: Kids Can Press, 2011.

Morgan, Allen. *Sadie and the Snowman.* Toronto: Kids Can Press, 1985.

Munsch, Robert. *Millicent and the Wind.* Toronto: Annick Press, 1984.

Na, Il Sung. *Snow Rabbit, Spring Rabbit: A Book of Changing Seasons.* New York: Knopf Books for Young Readers, 2011.

*Native Northwest. *Goodnight World: Animals of the Native Northwest.* Vancouver, BC: Native Northwest, 2013.

Nicholson, Caitlin. *niwîcihâw: I Help*. Toronto: Groundwood, 2008.

Orie, Sandra de Coteau. *Did You Hear the Wind Sing Your Name? An Oneida Song of Spring*. London, UK: Walker Children's, 1995.

*Robertson, David Alexander. *Warren Whistles at the Sky*. Winnipeg, MB: First Nations Education Resource Centre, 2016.

*Robertson, Joanne. *The Water Walker*. Toronto: Second Story Press, 2017

Rockwell, Anne. *Four Seasons Make a Year*. New York: Walker, 2004.

Sendak, Maurice. *Chicken Soup With Rice: A Book of Months*. New York: HarperCollins, 2014.

Shaw, Charles G. *It Looked Like Spilt Milk.* New York: HarperCollins, 1992.

Sidman, Joyce. *Red Sings from Treetops: A Year in Colours*. Boston: Houghton Mifflin Harcourt, 2009.

Silverstein, Shel. *The Giving Tree*. New York: HarperCollins, 2014.

Snedeker, Joseph. *The Everything Kids' Weather Book*. Avon, MA: Simon & Schuster, 2012.

Stewart, Melissa. *Droughts. Let's-Read-and-Find-Out Science 2*. New York: HarperCollins, 2017.

*Tharp-Thee, Sandy. *The Apple Tree*. Oklahoma City, OK: RoadRunner Press, 2015.

Vanasse, Deb. *Under Alaska's Midnight Sun*. Seattle: Sasquatch Books, 2005.

*Waboose, Jan Bourdeau. *Morning on the Lake*. Toronto: Kids Can Press, 1999.

*Waboose, Jan Bourdeau. *Skysisters*. Toronto: Kids Can Press, 2002.

Wilcox, John. *Charlie the Chinook*. Qualicum Beach, BC: Ravenrock Publishing, 1998.

▶

Portage & Main Press, 2019 · *Hands-On Science for British Columbia · Land, Water, and Sky for Grades K–2* · ISBN: 978-1-55379-797-5

*Williams, Maria. *How Raven Stole the Sun*. New York: Abbeville Kids, 2001.

Zolotow, Charlotte. *When the Wind Stops.* Toronto: HarperCollins Canada, 1997.

Websites

- **https://pbskids.org/caillou/games/ dresscaillou.html**
  PBS Kids—Caillou: Check the weather, and dress Caillou in the appropriate clothing for the weather.

- **www.theideabox.com**
  The Idea Box: Crafts, activities, music, and more for young children. Click on "Seasonal" to find activities, crafts, and projects with seasonal themes.

- **www.weatherwizkids.com**
  Weather Wiz Kids: Find weather experiments, activities, tools, and a variety of information related to daily and seasonal changes.

- **www.sheppardsoftware.com/ scienceforkids/seasons/seasons.htm**
  Sheppard Software—Seasons: Interactive website featuring a variety of activities for each of the four seasons.

- **www.native-languages.org/houses.htm**
  Native Languages of the Americas—Native American Houses: Information about Native American housing, featured on this website dedicated to the survival of Native American languages, particularly through the use of Internet technology.

- **exchange.smarttech.com/search?q=life+ cycle+sequence&carousel=true**
  SMART Exchange: Teacher Lesson Plans and resources submitted by educators from around the world, top educational publishers, and SMART. Enter relevant terms into the search engine to access a variety of activities that you can modify to suit the needs of

your students. Creating a SMART exchange account is free.

- **www.greenplanetfilms.org/product/ stories-of-the-seventh-fire-winter-spring/**
  Green Planet Films—Stories From the Seventh Fire: Winter and spring stories can be ordered from this website.

- **www.indigenouspeople.net/ chipewyn.htm**
  Athabasca Chipewyan First Nation (Dene/ Suline/Soline) Literature: This site links to Athabascan languages, stories, and more.

- **www.layers-of-learning.com/raven-stole- sun-native-american-raven-legend/**
  Layers of Learning: Go to this site for a free printable version of the raven story that students can illustrate.

- **http://publications.gc.ca/collections/ Collection/R72-278-2001E.pdf**
  Indigenous and Northern Affairs Canada— The Learning Circle: Classroom Activities on First Nations in Canada, Ages 8–11. See unit 3, "Water: Its Many Uses."

- **www.sciencekids.co.nz/sciencefacts/ energy/windenergy.html**
  Science Kids—Wind Energy Facts: Contains interesting wind-energy facts. Also, find experiments, games, projects on a wide range of science and technology topics.

- **water.usgs.gov/edu**
  US Geological Survey (USGS) Water Science School: This thorough site offers information on many aspects of water such as water cycles, water quality, and more, along with visuals, charts, maps, and an interactive centre.

- **response.restoration.noaa.gov/taxonomy/ term/87/all**
  National Oceanic and Atmospheric Administration (NOAA) Office of Response and Restoration—Exxon Valdez Oil Spill:

Portage & Main Press, 2019 · Hands-On Science for British Columbia · Land, Water, and Sky for Grades K–2 · ISBN: 978-1-55379-797-5

Includes information about immediate response to the spill, impact of the spill, and information about ongoing restoration and research activities, and more.

- **water.usgs.gov/edu/ watercyclesnowmelt.html**
  USGS Water Science School—Snowmelt: Find information about runoff from snowmelt, and how it contributes to the global movement of water.

- **https://water.usgs.gov/edu/watercycle.html**
  USGS Water Science School—The Water Cycle for Schools: Information aboutglobal water distribution and the water cycle, including a detailed diagram.

- **science.nasa.gov/earth-science/ oceanography/ocean-earth-system/ ocean-water-cycle**
  Water Cycle—NASA Science: Information about the water cycle.

- **https://www.canada.ca/en/health-canada/services/environmental-workplace-health/water-quality/drinking-water.html**
  Health Canada—Drinking water quality, contaminants and safety. How Canada enacts drinking water advisories.

- **www.regnery.com/books/ freddy-the-frogcaster/**
  Regnery Publishing—Freddy the Frogcaster: Free activity guides for Janice Dean's *Freddy the Frogcaster* books. Find three downloadable files: (1) Experiments Activity Guide; (2) Weather Journal; and (3) Weather Journal.

- **https://www.hellobc.com/plan-your-trip/ climate-weather/**
  HelloBC—Climate & Weather: Know what to expect when travelling in BC: Information about the weather and seasons in British Columbia.

- **https://www.crd.bc.ca/education/ our-environment/watersheds/ watershed-basics/water-cycle**
  Capital Regional District on Vancouver Island—Watershed Basics: Information about the water cycle and watershed. Includes diagrams and a list of important terms related to the water cycle.

- **http://www.sac-isc.gc.ca/eng/ 1506514143353/1533317130660**
  Indigenous Services Canada—Ending long-term drinking water advisories: Information on long-term drinking water advisories and steps the government is taking to end them.

- **www.fnha.ca/what-we-do/ environmental-health/drinking-water-advisories**
  First Nations Health Authority—Drinking Water Advisories: Information regarding drinking water advisories in First Nations Communities in British Columbia.

- **https://www.openschool.bc.ca/ elementary/my_seasonal_round/pdf/ SeasonalRound_unit.pdf**
  Open School BC: "My Seasonal Round: An Integrated Unit for Elementary Social Studies and Science."

- **https://vancouversun.com/news/local-news/more-than-half-of-b-c-s-school-districts-had-unsafe-lead-levels-in-drinking-water-sources-in-2016**
  Vancouver Sun: "More than half of B.C.'s school districts had unsafe lead levels in drinking water sources in 2016." Gordon Hoekstra and Lori Culbert.

- **http://secwepemc.sd73.bc.ca/sec_village/ sec_round.html**
  "Resource gathering: Seasonal Rounds Stsîllen." Information about how and when the Secwepemc gather resources.

Portage & Main Press, 2019 · *Hands-On Science for British Columbia · Land, Water, and Sky for Grades K–2* · ISBN: 978-1-55379-797-5

Portage & Main Press, 2019 · Hands-On Science for British Columbia · Land, Water, and Sky for Grades K–2 · ISBN: 978-1-55379-797-5

- **https://www.theguardian.com/world/2017/mar/16/new-zealand-river-granted-same-legal-rights-as-human-being**
  The Guardian: "New Zealand river granted same legal rights as human being." Eleanor Ainge Roy.

- **https://www.cbc.ca/news/canada/british-columbia/first-nations-water-solutions-1.3482568**
  CBC British Columbia: "Drinking Water on First Nations reserves an ongoing problem in B.C., says professor."

- **https://www.theglobeandmail.com/news/british-columbia/more-drinking-water-advisories-for-bc-than-any-other-province-report-finds/article23488553/**
  The Globe and Mail: "More drinking-water advisories for B.C. than any other province, report finds." Maura Forrest.

- **https://prometheanplanet.com/**
  Promethean Planet: Search for free calendar routines and other resources on this website.

- **http://appadvice.com/appguides/show/time-lapse-photography-apps**
  Time Lapse Photography Apps: There are several apps for tablets to choose from.

- **www.stevespanglerscience.com/store/uv-color-changing-beads.html**
  Steve Spangler Science: Purchase UV color changing beads on this site.

Videos

- **https://www.youtube.com/watch?v=dNo2hFTMay4**
  "When Raven Stole the Moon." Gishkishenh (1:51).

- **https://www.youtube.com/watch?v=VVloVe4jWMU**
  "Nocturnal Animal Noises." Katy Lange (2:21).

- **https://www.youtube.com/watch?v=yfNiIF3NWCM**
  "Nocturnal Animals (Night Animals)." Socratica Kids (1:49).

- **www.youtube.com/watch?v=FAnDIYRycqs**
  "The Life of Water. Water Which Gives Life–Water Project H20oooh!" UNESCO Venice Office. (2:08)

- **www.youtube.com/watch?v=al-do-HGulk&t=9s**
  "The Water Cycle." National Science Foundation. (6:46)

- **https://www.youtube.com/watch?v=74Y38Oy4AM4**
  "Raven Steals the Light." STORYHIVE (9:27).

- **https://www.youtube.com/watch?v=bzJ-mLcWiXs**
  "Winter Homes" SCESvids (3:35).

- **https://www.youtube.com/watch?v=H1GwODsJgSg**
  "Summer Home Materials" SCESvids (2:24)

- **https://www.youtube.com/watch?v=fVsONlc3OUY**
  "Night of the Northern Lights" Maciej Winiarczyk (2:22).

- **https://www.youtube.com/watch?v=fVMgnmi2D1w**
  "NASA UHD Video: Stunning Aurora Borealis from Space in Ultra-High Definition" Space Videos (4:36).

- **https://www.youtube.com/watch?v=iWjBjE4lys8**
  "Luke Howard – the man who named the clouds" OPALexplorenature (23:41).

Initiating Event: What Do We Observe, Think, and Wonder About Land, Water, and Sky?

1

Information for Teachers

In this lesson, students will participate in place-based learning to explore the land, water, and sky in a local natural environment. Plan ahead to select a location to ensure the area includes natural bodies of water. Also encourage students to offer suggestions of natural areas in your local region.

If possible, invite a local Elder or Knowledge Keeper to participate in this learning experience. They may be able to share relevant stories, as well as knowledge of the land, water, and sky.

NOTE: See Indigenous Perspectives and Knowledge, page 33, for guidelines for inviting Elders and Knowledge Keepers to speak to students.

Materials

- chart paper
- markers
- digital cameras
- magnifying glasses
- tweezers
- stretch gloves
- string or yarn
- garden tools (e.g., shovels, forks, trowels)
- Learning Centre Task Card: How Can I Sort Pictures of Land, Water, and Sky? (4.1.1)

Engage

Explain to students that they will be exploring a natural environment through a nature walk.

Discuss the location for this place-based learning. Ask:

- Who has been to this place before?
- How did you get there?
- What is it like there?
- What do you think we will see there? Smell? Hear? Feel?

Have students share their background knowledge, predictions, inferences, and ideas about the natural environment they will visit and the objects they might investigate there. Record their ideas on chart paper. Later, students can refer back to their ideas to see how their thinking changes.

Have students share with a partner what they are most excited about in visiting this location.

To help students develop a stronger image of their community and surrounding area, use a map of the local area to identify where the place-based learning location is in relation to the school. This is an excellent opportunity to identify the importance of place.

Identify on whose traditional territory the school is located, the traditional territory of the location for the place-based learning (if different), as well as the traditional names for both locations. The following map, "First Nations in British Columbia," from Indigenous Services Canada can be used for this purpose: <www.aadnc-aandc.gc.ca/DAM/DAM-INTER-BC/STAGING/texte-text/inacmp_1100100021016_eng.pdf>.

Incorporate land acknowledgment when students have learned on whose territory the school and place-based learning location are located. The following example can be used for guidance:

- We would like to acknowledge that we are gathered today on the traditional, ancestral, and unceded territory of the _____ people, in the place traditionally known as _____.

NOTE: Many school districts have established protocols for land acknowledgment. Check with colleagues who support Indigenous education to see if there are specific protocols to follow.

Portage & Main Press, 2019 · Hands-On Science for British Columbia · Land, Water, and Sky for Grades K–2 · ISBN: 978-1-55379-797-5

Review any other protocols for field trips, providing students with opportunities to ask questions and clarify expectations.

⚠️ **SAFETY NOTE:** Remind students never to taste anything without permission. Review other safety considerations such as plants that may cause skin irritations and bodies of water that may be present at the place-based learning location. Also be aware of students' allergies during the activity.

Discuss with students the importance of respecting nature. Have them brainstorm ways in which they can demonstrate this. Create an anchor chart on chart paper to display in the classroom. This can be reviewed each time students go outside to visit a natural environment. For example:

- Be respectful of all living things.
- Never break off branches from trees or pick wild flowers.
- Collect only a few objects to take back to the classroom.
- Always clean up after ourselves

Introduce the guided inquiry question: **What do we observe, think, and wonder about land, water, and sky?**

Explore Part One

Provide time for students to explore the area freely (under adult supervision). Provide access to materials such as digital cameras, magnifying glasses, tweezers, stretch gloves, and garden tools for exploration. As students explore, pose questions for them to ponder. For example:

- What are you examining?
- Why is it interesting to you?
- What can you tell me about it?
- What do you wonder about it?
- What do you see? Feel? Smell? Hear?

Encourage students to share their observations, thinking, and questions with you, their peers, and other supervisors.

Regroup to discuss this initial exploration. Reflect on the predictions students made before the trip. Ask:

- What things did you think you would see? Smell? Hear? Feel?
- How were your predictions? Did you see, hear, smell, and hear what you thought you would?
- What surprised you?

Have students share their observations, ideas, and questions.

Explore Part Two

Have students work in groups and choose one area of mutual interest to explore again. Groups can use string to create a border so they can zoom in and focus on one small area. Have them explore this area to learn more about the land and its features.

NOTE: For more information on this learning experience, see *A Place for Wonder,* by Georgia Heard (see the lesson "One Small Square").

Encourage the groups to share with each other their observations, ideas, and questions as they conduct deeper exploration.

Regroup to discuss this group exploration. Have each group show the rest of the class their chosen area and then share their observations, ideas, and questions. Guide the discussion with the Observe, Think, Wonder format:

- What did you observe in your special place?
- What did you think about the place?
- What did you wonder about the place?

Encourage other students to share their observations, ideas, and questions as well.

Portage & Main Press, 2019 · Hands-On Science for British Columbia · Land, Water, and Sky for Grades K–2 · ISBN: 978-1-55379-797-5

Explore Part Three

Gather students together and have them lie on their backs and look at the sky.

 SAFETY NOTE: Remind students never to look directly at the Sun.

Have students share their observations of what they see in the sky.

Next, have students work in pairs and choose a location to lie down and sky-gaze. Have pairs share their observations of the sky's features.

Regroup and review what students have learned about the sky. Ask:

- How would you describe the sky?
- What kinds of things did you see in the sky?
- Did you see anything moving in the sky?
- What colours did you see?
- What sounds did you hear coming from the sky?

Have students share their observations.

Explore Part Four

As a class, observe water in the area.

 SAFETY NOTE: Decisions on how best to observe water will depend on access to adult supervision and water features.

Discuss the features of the water itself. Brainstorm words to describe the water's appearance (e.g., how it looks, how it moves). If accessible and safe, have students touch and smell the water (or collect a water sample in a pail for students to touch and smell). Brainstorm words to describe how the water feels and smells.

Explore Part Five

Organize the class into working groups, and provide each group with a digital camera.

NOTE: This task may work best if a supervisor/ adult is with each group.

Have groups choose one area of mutual interest to explore and take photographs of land, water, and sky. Encourage students to share with each other their observations, ideas, and questions as they conduct deeper exploration.

Explore Part Six

Back in the classroom, regroup to discuss this group exploration. Have each group show the rest of the class their photographs, and then share observations, ideas, and questions.

Conclude the discussion with the Observe, Think, Wonder format:

- What did you observe about the land, water, and sky?
- What did you think about the land, water, and sky?
- What did you wonder about the land, water, and sky?

Encourage other students to share their observations, ideas, and questions as well.

Expand

To prepare students for journaling, complete a journal entry together as a class. This will require a portable whiteboard, or chart paper with a sturdy backboard for writing.

As a class, choose a place to sit and journal. Brainstorm journaling ideas. At the kindergarten to grade-two level, journaling can involve many different activities. Students may want to:

- sketch or colour the land, water, or sky they see around them
- sketch flowers or leaves
- identify emotions, recording feelings about the land, water, or sky (students may use happy faces, emojis, or their own designs for emotions)

▶

Portage & Main Press, 2019 · Hands-On Science for British Columbia · Land, Water, and Sky for Grades K–2 · ISBN: 978-1-55379-797-5

Portage & Main Press, 2019 · *Hands-On Science for British Columbia · Land, Water, and Sky for Grades K–2* · ISBN: 978-1-55379-797-5

■ record the sounds they hear, using pictures and words

■ record the movements they see, using pictures and words

■ photograph the land, water, or sky and sketch it

■ draw labelled diagrams showing the land, water, or sky

■ draw a map of the local area or journaling location

■ lie back and observe the sky to sketch what they see moving

■ use all their senses to describe the land, water, or sky using pictures and/or words

■ write a poem about what they observe or feel about the land, water, or sky

After working through a few journaling activities together, distribute place-based journals and supplies to students. Have them choose a place where they would like to sit and journal, and have them choose one journaling strategy to record.

Formative Assessment

Photograph students as they journal to collect as evidence of learning activities. Be sure to document students' thinking after journaling. Meet with students individually to have them share their thoughts about their observations of objects and materials in nature, the journaling experience, and what they recorded in their journal. Focus on their oral and written communication as they express and reflect on personal experiences of place. Use photographs taken as they journal to inspire reflection. Use the INDIVIDUAL STUDENT OBSERVATIONS template on page 51 to record interview highlights. Provide descriptive feedback to students about how they express and reflect on personal experiences of place.

Learning Centre

At the learning centre, provide the collection of photographs taken during the place-based

learning activity and a copy of the Learning-Centre Task Card: How Can I Sort Pictures of Land, Water, and Sky (4.1.1):

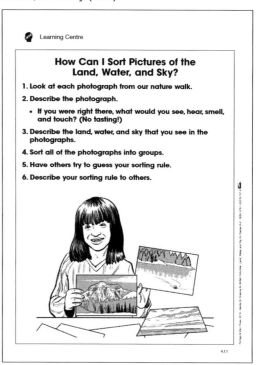

Download this template at <www.portageandmainpress.com/product/HOSLandWaterSkyK2>.

Have students work in groups to examine the photographs and describe them to peers. Then have them sort the photographs according to their own rules. Exploration of these photographs may inspire further inquiry. Encourage students to ask questions and find ways to answer those questions, either at the centre or as personalized learning experiences.

Embed Part One: Talking Circle

NOTE: See page 16 for more information about Talking Circles.

Revisit the guided inquiry question: **What do we observe, think, and wonder about land, water,**

and sky? Have students share their experiences and knowledge, provide examples, and ask further inquiry questions.

Embed Part Two

- Focus on students' use of the Core Competencies. Have students reflect on how they used one of the Core Competencies (Thinking, Communicating, or Personal and Social Skills) during the various lesson activities. Project one of the CORE COMPETENCY DISCUSSION PROMPTS templates (pages 38–42), and use it to inspire group reflection. Referring to the template, choose one or two "I Can" statements on which to focus. Students then use the "I Can" statements to provide evidence of how they demonstrated that competency. Ask questions directly related to that competency to inspire discussion. For example:
  - Where did you get your ideas for your place-based journal entry today? (Creative Thinking)

  Have students reflect orally, encouraging participation, questions, and the sharing of evidence. (See page 29 for more information on these templates.)

  As part of this process, students can also set goals. For example, ask:
  - What would you do differently next time and why?
  - How will you know if you are successful in meeting your goal?
- To encourage self-reflection, provide prompts that students can use to cite examples of how they have used the Core Competencies in their learning. For this purpose, the CORE COMPETENCY SELF-REFLECTION FRAMES (pages 43–47) can be used throughout the learning process. There are five frames provided to address the Core Competencies: Communication, Creative Thinking, Critical Thinking, Positive Personal and Cultural Identity, and Personal Awareness and Responsibility. Teachers can conference individually with students to support self-reflection, or students may complete prompts using words and pictures.

  Again, have students set goals by considering what they might do differently on future tasks and how they will know if they are successful in meeting their goal.

NOTE: Use the same prompts from these templates over time to see how thinking changes with different activities.

Enhance

- **Family Connections:** Provide students with the following sentence starters:
  - When we go for a walk outdoors, we notice _____ about the land.
  - We notice _____ about water.
  - We notice _____ about the sky.

  Have students complete the sentence starters at home. Family members can help students draw and write about this topic. Have students share their sentences with the class.
- Follow up the place-based learning activities in this lesson with a class discussion about the experience. Construct an Observe, Think, Wonder chart as in the following example:

Observe	Think	Wonder
		?

- Have students examine photographs from the place-based learning experience, share their ideas, and record these in the appropriate columns. Also have them compare observations to predictions discussed before to the place-based learning experience.

Portage & Main Press, 2019 · Hands-On Science for British Columbia · Land, Water, and Sky for Grades K–2 · ISBN: 978-1-55379-797-5

2 What Can We Learn About Land, Water, and Sky Through Storytelling?

Information for Teachers

Storytelling makes learning more meaningful and engaging for students by connecting science concepts to real life and to other subjects. In this lesson, students will participate in a variety of storytelling experiences related to land, water, and sky.

Plan ahead to invite guest storytellers. Also encourage students to offer suggestions for storytellers from your local area, including family and community members.

If possible, invite a local Elder or Knowledge Keeper to participate in this learning experience They may be able to share relevant stories about land, water, and/or sky.

NOTE: See Indigenous Perspectives and Knowledge, page 9, for guidelines for inviting Elders and Knowledge Keepers to speak to students.

Introduce students to new books with a book walk (see Engage). Use picture books related to objects and materials to provide students with opportunities to build background, enhance interest, generate questions, make predictions, examine visuals, learn new concepts, and share their growing knowledge with others.

Materials

- picture books about land, water, and sky including books by Indigenous authors, and books reflective of the cultural make-up of the class and community (one book for each pair of students)
- Template: Interview Guide: Sharing Stories About Land, Water, and Sky (4.2.1)
- Learning-Centre Task Card: What Can I Learn About Land, Water, and Sky? (4.2.2)
- drawing paper
- art supplies

Engage

Model oral storytelling by sharing a short story about land, water, and/or sky. This might be a family story, personal experience, or a story passed down from others. Consider telling a story about:

- a family hike
- swimming or boating
- star-gazing or a trip to the planetarium

Encourage students to ask questions and share ideas after the oral storytelling.

Have students share a short story about land, water, and/or sky with a partner.

Introduce the guided inquiry question: **What can we learn about land, water, and sky through storytelling?**

Explore Part One

Select one of the picture books about land, water, and/or sky. Use the book to model a book walk with the following steps:

1. Show students the cover of the picture book.
2. Discuss the cover illustration.
3. Point out and discuss the various features of the cover, including the title, author, and illustrator.
4. Have students predict what the book is going to be about, based on their observations of the book's cover.
5. Walk through the pages, discussing the pictures but not reading the text. Ask questions about the illustrations. Have students share what they think is happening on each page.
6. Discuss the final illustrations, and have students predict how the story ends.
7. Discuss the sequence of events in the story.
8. Have students share how the book makes them feel, as well as what they wonder about.

Portage & Main Press, 2019 · Hands-On Science for British Columbia · Land, Water, and Sky for Grades K–2 · ISBN: 978-1-55379-797-5

2

9. Have students share stories and experiences related to what they saw in the book.

After students have shared their ideas and experiences, read the book to the class.

Explore Part Two

Organize the class into pairs and have each pair choose one of the picture books about land, water, and/or sky. Have pairs do a book walk together, as modelled above.

When they have completed the book walk, have each pair introduce their book to the class, presenting the pictures and their ideas about the story.

Read the books aloud to the class. Focus on the features of land, water, and/or sky, as depicted in the books.

Include these books at the Learning Centre Library.

As books are shared and discussed as a class, consider using a Talking Circle to provide opportunities for students to communicate their thoughts, ideas, and questions about the books. (See page 16 in the for more information about Talking Circles).

Explore Part Three

Many cultures pass down knowledge and history through oral storytelling. Provide an opportunity for students to benefit from these traditions to learn about land, water, and sky.

Invite a local Elder or Knowledge Keeper to share stories with the class.

Also consider inviting storytellers from other cultures, reflective of your local community, to share their oral traditions with the class.

NOTE: It is important to prepare for guest speakers, and to ensure students are appropriately prepared as well. Review behavioural expectations, and discuss questions that students may wish to ask the guest. Be sure to have students thank the speaker for the visit, and consider following up with written or illustrated thank-you notes. It is also important to consider protocols for Elders. Please see the *Science First Peoples: Teacher Resource Guide* (First Nations Schools Association, 2016) for guidelines and considerations.

Explore Part Four

Have students interview family members and ask them to share stories about land, water, and/or sky. Ensure students know to clearly communicate with family members that the stories will be shared publicly. Review that First Peoples Principles of Learning involves recognizing that some knowledge is sacred and only shared with permission and/or in certain situations (please see page 12 in for more information on First Peoples Principles of Learning).

Story themes might include:

- a visit to the mountains, prairies, or other geographically unique land regions
- land and water in the community
- land, water, and/or sky observed on a camping trip, vacation, or canoe trip
- land, water, and/or sky observed on a hunting or fishing trip
- land, water, and/or sky from other countries
- family or community gardens
- annual berry picking trips
- sky-gazing, astronomy, or experiences with telescopes
- sky feature experiences such as northern lights, rainbows, shooting stars, eclipses

Students may use the Template: Interview Guide: Sharing Stories About Land, Water, and Sky (4.2.1):

▶

Portage & Main Press, 2019 · *Hands-On Science for British Columbia · Land, Water, and Sky for Grades K–2* · ISBN: 978-1-55379-797-5

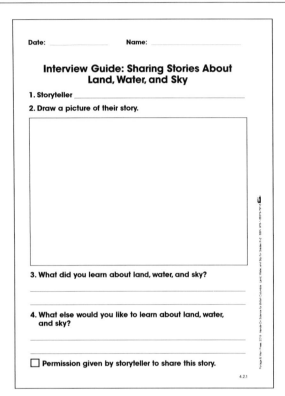

Portage & Main Press, 2019 · *Hands-On Science for British Columbia · Land, Water, and Sky for Grades K–2* · ISBN: 978-1-55379-797-5

Download this template at <www.portageandmainpress.com/product/HOSLandWaterSkyK2>.

NOTE: If using this template, review the activity before students take it home, so they are familiar with the questions. Family members may help students complete the template.

Students may also use other formats for recording these stories, including audio recordings, illustrations, or using oral sharing techniques to retell the family story.

After students complete this task at home, have them share stories with the class. During discussion, have students do the following:

- Tell the class about the person they interviewed.
- Provide a summary of the story.
- Explain how they know that the story was about land, water, and/or sky.

NOTE: Based on these interviews, students may be interested in inviting guests to share other stories about land, water, and/or sky. Family members, Elders and Knowledge Keepers, school staff, community members, and staff of local organizations who work with land, water, and/or sky may be accessed to share stories about living things.

Expand

Provide students with an opportunity to explore oral storytelling about land, water, and sky further by posing their own questions for personalized inquiry. They may wish to:

- Initiate a project at the Makerspace, such as designing and constructing a puppet theatre with puppets to recreate a story they have heard or made up themselves.
- Explore Loose Parts bins related to land, water, and/or sky with collections of objects found on the land (e.g., rocks, leaves, pine cones, seeds, grasses, soil samples). Students might use the Loose Parts to represent what they know, can do, and/or understand about the land, water, and sky. Using Loose Parts, students may create a picture depicting the land, water, and/or sky on an empty picture frame. They can then tell a story of something that could have happened in their scene. Have students photograph their creation to document their learning.
- Use water and sand tables. Include a variety of materials for students to explore properties of sand and water. Provide an opportunity for storytelling during the exploration.
- Make up their own oral stories about land, water, and/or sky to share with the class.
- Design and create models of characters from stories they have heard.
- Conduct an investigation based on their own inquiry questions.

As students explore and select ideas to expand learning, provide support and guidance as needed, and offer access to materials and resources that will enable students to conduct their chosen investigations.

Learning Centre

NOTE: In preparation for the learning centre library, complete this activity as a class to make it more successful. As a class, explore and discuss questions such as:

- How do we read pictures?
- How do we find answers to our inquiry questions when we cannot read all the words in a book?

At the learning centre, set up a classroom Land, Water, and Sky Library. Provide a variety of picture books on the topic, as well as art or drawing paper and art supplies. Also, provide a copy of the Learning-Centre Task Card: What Can I Learn About Land, Water, and Sky? (4.2.2):

> Learning Centre
>
> ### What Can I Learn About Land, Water, and Sky?
>
> 1. Choose a book.
> 2. Look at the cover of the book.
> 3. Discuss the cover illustration.
> 4. Find the title, author, and illustrator.
> 5. Predict what the book is going to be about, based on the book's cover.
> 6. Walk through the pages, discussing the pictures but not reading the text.
> 7. Read the book.
> 8. Draw a picture to show what you learned about land, water, and sky.
>
> 4.2.2

Download this template at <www.portageandmainpress.com/product/HOSLandWaterSkyK2>.

Have students choose anything they want to read from the Library.

NOTE: Teachers may choose to focus on one topic (either land, water, or sky) and display books related to only that theme. Alternately, have a collection of books on all three topics.

Encourage students to conduct book walks, examine and discuss illustrations in the books, and share their ideas with peers. Students can then draw a picture reflective of their ideas.

NOTE: Depending on students' literacy skills, some may choose to print the title and author of a book and create an illustration from that source.

Display students' work in the Land, Water, and Sky Library for everyone to enjoy.

Embed Part One: Talking Circle

Revisit the guided inquiry question: **What can we learn about land, water, and sky through storytelling?** Have students share their experiences and knowledge, provide examples, and ask further inquiry questions.

Embed Part Two

- Focus on students' use of the Core Competencies. Have students reflect on how they used one of the Core Competencies (Thinking, Communicating, or Personal and Social Skills) during the various lesson activities. Project one of the CORE COMPETENCY DISCUSSION PROMPTS templates (pages 38–42), and use it to inspire group reflection. Referring to the template, choose one or two "I Can" statements on which to focus. Students then use the "I Can" statements to provide evidence of how they demonstrated that competency. Ask

▶

Portage & Main Press, 2019 · *Hands-On Science for British Columbia · Land, Water, and Sky for Grades K–2* · ISBN: 978-1-55379-797-5

Portage & Main Press, 2019 · Hands-On Science for British Columbia · Land, Water, and Sky for Grades K–2 · ISBN: 978-1-55379-797-5

questions directly related to that competency to inspire discussion. For example:

■ How did you share your learning today? (Communication)

Have students reflect orally, encouraging participation, questions, and the sharing of evidence. (See page 29 for more information on these templates.)

As part of this process, students can also set goals. For example, ask:

■ What would you do differently next time and why?

■ How will you know if you are successful in meeting your goal?

■ To encourage self-reflection, provide prompts that students can use to cite examples of how they have used the Core Competencies in their learning. For this purpose, the CORE COMPETENCY SELF-REFLECTION FRAMES (pages 43–47) can be used throughout the learning process. There are five frames provided to address the Core Competencies: Communication, Creative Thinking, Critical Thinking, Positive Personal and Cultural Identity, and Personal Awareness and Responsibility. Teachers can conference individually with students to support self-reflection, or students may complete prompts using words and pictures.

Again, have students set goals by considering what they might do differently on future tasks and how they will know if they are successful in meeting their goal.

NOTE: Use the same prompts from these templates over time to see how thinking changes with different activities.

3 | What Do We Know About Land, Water, and Sky?

Materials

- three different colours of sticky notes
- chart paper
- markers
- digital cameras (or clipboards, drawing paper, and pencils)

Engage

Ask students to think about what they know about land, water, and/or sky. Using one colour of sticky note, have students record one thing they know, using pictures or text.

NOTE: Save these sticky notes for the next activity.

Conduct a Think, Pair, Share activity. Have each student share their idea with an elbow partner. Next, for class sharing, have each student role-play or mime one thing they know about land, water, and/or sky. Model this for the class using your own ideas (e.g., demonstrate digging in the garden or climbing a mountain to show land, swimming or kayaking to show water, or flying a kite to show sky). Challenge students to guess the action, then have students take turns role-playing while their classmates guess.

Introduce the guided inquiry question: **What do we know about land, water, and sky?**

Explore Part One

On chart paper or mural paper, create a large concept web to record ideas, as in the following example:

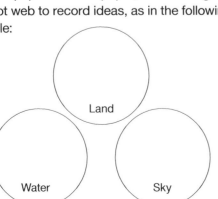

Display the sticky notes from the Engage activity, and have students sort them onto the concept web. Ask:

- What questions do you have about land, water, and sky?
- What would you like to learn about land, water, and sky?

Record students' questions on a different colour of sticky note. Attach the notes onto the concept web, on the outside edges of the appropriate circle.

Refer to the concept web throughout the module. Add answers to the students' questions to the web, outside each circle, with a different colour of sticky note, as students gather new information and acquire new knowledge. As students generate more inquiry questions, these may be added as well.

Explore Part Two

Begin a word wall to display new terms introduced throughout the module, as well as illustrations, photographs, and examples. When possible, add terminology in languages other than English on the word wall, including Indigenous languages. This is a way of acknowledging and respecting students' cultural backgrounds, while enhancing learning for all students.

Word	Picture	Example
Sun hlo<u>k</u>s (Nisga'a language)		The Sun gives us light and heat.

▶

Portage & Main Press, 2019 · *Hands-On Science for British Columbia · Land, Water, and Sky for Grades K–2* · ISBN: 978-1-55379-797-5

Portage & Main Press, 2019 · *Hands-On Science for British Columbia · Land, Water, and Sky for Grades K–2* · ISBN: 978-1-55379-797-5

3

NOTE: A variety of online dictionaries may be used as a source for translations. For example:

- <http://www.freelang.net/online/haida.php>
- <http://www.firstvoices.com/en/Halqemeylem>

Online dictionaries are also available for languages that may be reflective of the class cultural makeup.

As a class, discuss important words students have been using and learning (e.g., the Sun). Record these terms and have students make drawings on large post-it notes, and label the parts of their drawings. Their drawings can be added to the word wall in the centre column, along with photographs and other pictures collected throughout the module.

Expand

Provide students with an opportunity to explore land, water, and sky further by posing their own questions for individualized inquiry. They may wish to:

- Initiate a project at the Makerspace, such as designing and constructing a model that illustrates land, water, and sky (e.g., a mountain, waterfall, cloudy sky).
- Explore Loose Parts bins related to land, water, and/or sky with collections of natural objects that float in water (e.g., seed pods, pine cones, sticks).
- Make and display a collection of rocks and minerals.
- Explore sinking and floating with a variety of objects from the classroom in the water table. Students can design ways to record their observations.
- Use the sand table to explore what happens when water is added to sand. Students may explore what kinds of creations they can make with this new texture.
- Build a structure that floats in water.
- Conduct an investigation or experiment based on their own inquiry questions.

As students explore and select ideas to expand learning, provide support and guidance as needed, and offer access to materials and resources that will enable students to conduct their chosen investigations.

Embed Part One: Talking Circle

Revisit the guided inquiry question: **What do we know about land, water, and sky?** Have students share their experiences and knowledge, provide examples, and ask further inquiry questions.

Embed Part Two

- Add to the concept web as students learn new concepts, answer some of their own inquiry questions, and ask new inquiry questions.
- Focus on students' use of the Core Competencies. Have students reflect on how they used one of the Core Competencies (Thinking, Communicating, or Personal and Social Skills) during the various lesson activities. Project one of the CORE COMPETENCY DISCUSSION PROMPTS templates (pages 38–42), and use it to inspire group reflection. Referring to the template, choose one or two "I Can" statements on which to focus. Students then use the "I Can" statements to provide evidence of how they demonstrated that competency. Ask questions directly related to that competency to inspire discussion. For example:
  - How did you decide which questions to ask today?" (Critical Thinking)

Have students reflect orally, encouraging participation, questions, and the sharing of evidence. (See page 29 for more information on these templates.)

As part of this process, students can also set goals. For example, ask:

- What would you do differently next time and why?

Hands-On Science for British Columbia

- How will you know if you are successful in meeting your goal?

To encourage self-reflection, provide prompts that students can use to cite examples of how they have used the Core Competencies in their learning. For this purpose, the CORE COMPETENCY SELF-REFLECTION FRAMES (pages 43–47) can be used throughout the learning process. There are five frames provided to address the Core Competencies: Communication, Creative Thinking, Critical Thinking, Positive Personal and Cultural Identity, and Personal Awareness and Responsibility. Teachers can conference individually with students to support self-reflection, or students may complete prompts using words and pictures.

Again, have students set goals by considering what they might do differently on future tasks and how they will know if they are successful in meeting their goal.

NOTE: Use the same prompts from these templates over time to see how thinking changes with different activities.

Enhance

- **Family Connections**: Provide students with the following sentence starter:
  - A favourite body of water for our family is _____ because _____.

  Have students complete the sentence starter at home. Family members can help students draw and write about this topic. Have students share their sentences with the class.

Portage & Main Press, 2019 · *Hands-On Science for British Columbia · Land, Water, and Sky for Grades K–2* · ISBN: 978-1-55379-797-5

4 | What Are Some Differences Between Daytime and Nighttime?

Materials

- Picture Cards: Daytime Activities (4.4.1)
- Picture Cards: Nighttime Activities (4.4.2)
- chart paper
- markers
- scissors
- *Night and Day* by David Bouchard (also available on CD)
- interlocking cubes of different colours
- books that explore concepts related to daytime and nighttime (see Explore Part Three)
- Learning-Centre Task Card: What If...? (4.4.3)
- art paper
- pencil crayons
- large sheet of black construction paper
- large sheet of yellow construction paper
- clear tape
- sticky notes
- concept web (from lesson 3)

Engage

Create a sorting circle by cutting out a large half circle from each piece of construction paper. Tape the half circles together to create the sorting circle.

Print and cut out Picture Cards: Daytime Activities (4.4.1) and Picture Cards: Nighttime Activities (4.4.2):

Daytime Activities

4.4.1

Nighttime Activities

4.4.2

Portage & Main Press, 2019 · *Hands-On Science for British Columbia · Land, Water, and Sky for Grades K–2* · ISBN: 978-1-55379-797-5

4

Download these templates at <www.portageandmainpress.com/product/HOSLandWaterSkyK2>.

Put the sorting circle on the floor, and have students sit in a circle around it. Place one picture card from Daytime Activities (4.4.1) in the yellow section of the sorting circle. Then, place one picture card from Nighttime Activities (4.4.2) in the black section of the sorting circle. Ask:

- What do you think my sorting rule is?
- What are your clues?

Add another picture to each half circle. Ask:

- Why do you think these pictures are grouped together on the yellow paper?
- How are they the same?
- Why do you think these pictures are grouped together on the black paper?
- How are they the same?
- What is my sorting rule?

Give students an opportunity to share and compare their ideas and to sort the remaining pictures according to the daytime/nighttime sorting rule. Encourage students to suggest other activities.

Distribute sticky notes to each student to draw an activity connected to their own lives. Have each student show their illustration to the class, sort them, and place their illustration on the appropriate half of the circle.

Introduce the guided inquiry question: **What are some differences between daytime and nighttime?**

Explore Part One

Introduce students to Métis writer David Bouchard's storytelling with his children's book, *Night and Day* (also available on CD-ROM so pages can be projected). The story describes the different behaviours of various animal characters throughout the daytime and nighttime.

After reading the book, ask:

- Which animals are most active during the daytime?
- Which animals are most active during the nighttime?

As a class, create a T-chart and brainstorm a list of animals that are active in daytime and those that are active in nighttime. For example:

Animals Active in Daytime	Animals Active in Nighttime
robins	skunks
squirrels	owls
eagles	bats

Have students conduct further research about nocturnal and diurnal animals. Have students predict some useful features of nocturnal animals (e.g., owls have large eyes that let in as much light as possible, which allows them to see much better in the dark than most animals; skunks, foxes, and raccoons have good hearing to hear predators they cannot see in the dark).

Extend this activity to include Indigenous stories about local nocturnal and diurnal animals. For example:

- *Owls See Clearly At Night (Lii Yiiboo Nayaapiwak Lii Swer): A Michif Alphabet (L'Alfabet Di Michif)* by Julie Flett
- *Goodnight World* by Native Northwest
- *A Walk on the Shoreline* by Rebecca Hainnu
- *Strong Stories Coast Salish: Why Ravens and Wolves Hunt Together* by Celestine Aleck

Explore Part Two

Continue discussion about daytime and nighttime. Emphasize the presence of the Sun during the daytime and the absence of the Sun during the nighttime. Ask students:

Portage & Main Press, 2019 · Hands-On Science for British Columbia · Land, Water, and Sky for Grades K–2 · ISBN: 978-1-55379-797-5

4

- What changes when the Sun is not seen?
- What changes when it is visible again?

Divide a sheet of chart paper into two columns. Title the columns "Daytime" and "Nighttime." During the discussion, record students' ideas on the chart. Focus on different daytime and nighttime observations and activities, including the following:

- sights and sounds people may see or hear in the daytime and in the nighttime
- jobs people do in the daytime and jobs people do in the nighttime
- animals active during the daytime (diurnal) and animals active during the nighttime (nocturnal)

Encourage students to identify similarities and differences between daytime and nighttime, exploring events common to both day and night, as well as those that are different.

Discuss that the daytime/nighttime cycle is a repeating pattern (ABAB). Have students use interlocking cubes of different colours to represent this pattern. Have them choose two colours they think symbolize daytime and nighttime (e.g., yellow and black, or any combination based on their ideas) and create the pattern with the blocks.

Explore Part Three

Read a variety of books that explore concepts related to daytime and nighttime. For example:

- books related to the long hours of daylight in northern Canada such as *Arctic Lights, Arctic Nights* by Debbie S. Miller or *Under Alaska's Midnight Sun* by Deb Vanasse
- nonfiction books that focus on animal and plant behaviour both during the daytime and the nighttime (e.g., the behaviour of diurnal and nocturnal animals, and changes in

certain plants and flowers between day and night)
- *Goodnight Moon* by Margaret Wise Brown, which focuses on the changes that occur as nighttime falls

During the reading of each book, discuss the similarities and differences between daytime and nighttime.

Expand

Provide students with an opportunity to explore the differences between daytime and nighttime further by posing their own questions for individualized inquiry. They may wish to:

- Initiate a project at the Makerspace, such as making a model of a nocturnal animal in its habitat.
- Research a nocturnal animal (see page 26 for more information about inquiry through research).
- Create a picture book showing activities students do during the daytime and during the nighttime. Have students present their books during sharing time.
- Research a lunar or solar eclipse.
- Research the aurora borealis.
- Conduct an investigation based on their own inquiry questions.

As students explore and select ideas to expand learning, provide support and guidance as needed, and offer access to materials and resources that will enable students to conduct their chosen investigations.

Learning Centre

At the learning centre, provide art paper, pencil crayons, and a copy of the Learning-Centre Task Card: What If...? (4.4.3):

Portage & Main Press, 2019 · Hands-On Science for British Columbia · Land, Water, and Sky for Grades K–2 · ISBN: 978-1-55379-797-5

4

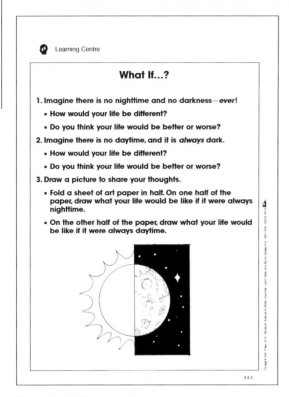

Learning Centre

What If...?

1. Imagine there is no nighttime and no darkness—*ever*!
   - How would your life be different?
   - Do you think your life would be better or worse?
2. Imagine there is no daytime, and it is *always* dark.
   - How would your life be different?
   - Do you think your life would be better or worse?
3. Draw a picture to share your thoughts.
   - Fold a sheet of art paper in half. On one half of the paper, draw what your life would be like if it were always nighttime.
   - On the other half of the paper, draw what your life would be like if it were always daytime.

4.4.3

Download this template at
<www.portageandmainpress.com/product/
HOSLandWaterSkyK2>.

Have students imagine what their lives could be like if there were no daytime or no nighttime. Ask them to use pictures and or words to record their reflections on the paper.

Embed Part One: Talking Circle

Revisit the guided inquiry question: **What are some differences between daytime and nighttime?** Have students share their experiences and knowledge, provide examples, and ask further inquiry questions.

Embed Part Two

- Add to the concept web as students learn new concepts, answer some of their own inquiry questions, and ask new inquiry questions.

- Add new terms and illustrations to the word wall. Include the words in languages other than English, as appropriate.

- Focus on students' use of the Core Competencies. Have students reflect on how they used one of the Core Competencies (Thinking, Communicating, or Personal and Social Skills) during the various lesson activities. Project one of the CORE COMPETENCY DISCUSSION PROMPTS templates (pages 38–42), and use it to inspire group reflection. Referring to the template, choose one or two "I Can" statements on which to focus. Students then use the "I Can" statements to provide evidence of how they demonstrated that competency. Ask questions directly related to that competency to inspire discussion. For example:

  - How did you grow as a learner today? (Positive Personal and Cultural Identity)

  Have students reflect orally, encouraging participation, questions, and the sharing of evidence. (See page 29 for more information on these templates.)

  As part of this process, students can also set goals. For example, ask:

  - What would you do differently next time and why?

  - How will you know if you are successful in meeting your goal?

- To encourage self-reflection, provide prompts that students can use to cite examples of how they have used the Core Competencies in their learning. For this purpose, the CORE COMPETENCY SELF-REFLECTION FRAMES (pages 43–47) can be used throughout the learning process. There are five frames provided to address the Core Competencies: Communication, Creative Thinking, Critical Thinking, Positive Personal and Cultural Identity, and Personal Awareness and Responsibility. Teachers can conference individually with students to support self-

Portage & Main Press, 2019 · *Hands-On Science for British Columbia · Land, Water, and Sky for Grades K–2* · ISBN: 978-1-5379-797-5

4

reflection, or students may complete prompts using words and pictures. Again, have students set goals by considering what they might do differently on future tasks and how they will know if they are successful in meeting their goal.

NOTE: Use the same prompts from these templates over time to see how thinking changes with different activities.

Enhance

- **Family Connections**: Provide students with the following sentence starter:
  - Favourite family activities during the daytime and nighttime are _____.

  Have students complete the sentence starter at home. Family members can help students draw and write about this topic. Have students share their sentences with the class.

- Have students explore why some animals stay up at night by showing them the following videos:
  - "Nocturnal Animal Noises." Go to <https://www.youtube.com/watch?v=VVloVe4jWMU>
  - "Nocturnal Animals (Night Animals)." Go to <https://www.youtube.com/watch?v=yfNilF3NWCM>

- Distribute to each student one whole paper plate and one half paper plate. On the whole paper plate, have students draw pictures of things that occur in the daytime (e.g., the Sun shines, children are at school or playing outside). On the half paper plate, have students draw pictures that show things that happen in the nighttime (e.g., the moon, the stars, and nocturnal animals come out). When their pictures are complete, have students use brass paper fasteners to attach the half plate over the whole plate, making certain the half plate is movable (see below). Students can now use their models to discuss daytime and nighttime activities.

Portage & Main Press, 2019 · Hands-On Science for British Columbia · Land, Water, and Sky for Grades K–2 · ISBN: 978-1-55379-797-5

5 | How Are the Months of the Year the Same and Different?

Materials

- calendars (collect, and encourage students to collect, from local businesses)
- large monthly calendar (commercial or teacher-made, paper or electronic, projected)
- chart paper
- markers
- Picture Cards: Weather and Special Events (4.5.1)
- glue stick (optional)
- scissors
- projection device (optional)
- *Chicken Soup With Rice: A Book of Months* by Maurice Sendak
- *A Canadian Year: Twelve Months in the Life of Canada's Kids* by Tania McCartney
- recycled scrap paper
- computer/tablet with internet access
- Learning-Centre Task Card: Acting Out the Months (4.5.2)
- concept web (from lesson 3)

Engage

Teach students the words and actions to "The Montherena." The tune and actions are the same as those for "The Macarena."

January (*right hand faces down*)

February (*left hand faces down*)

March (*right hand faces up*)

April (*left hand faces up*)

May (*right hand on left shoulder*)

June (*left hand on right shoulder*)

July (*right hand on right ear*)

August (*left hand on left ear*)

September (*right hand on left waist*)

October (*left hand on right waist*)

November (*right hand on right hip*)

December (*left hand on left hip*)

OH MONTHERENA! (*jump, and turn ¼ turn*)

(*repeat verse*)

Introduce the guided inquiry question: **How are the months of the year the same and different?**

Explore Part One

Organize the class into working groups. Provide each group with several sample calendars. Have students identify similarities and differences between the calendars.

Bring the class together again to debrief and share ideas. Ask:

- What did you notice about the different calendars?
- Do all the calendar pages look the same?
- How are they similar?
- How are they different?
- At which month does the calendar start?
- At which month does it end?
- How many months are in the calendar?
- When 12 months have passed, how much time has gone by?
- What is the sequence of months in a year?

Practise "The Montherena" song again, and record the months of the year on chart paper, in sequence. As a class, discuss the patterns students observe in the spelling, syllables, and sounds of the months.

Portage & Main Press, 2019 · *Hands-On Science for British Columbia · Land, Water, and Sky for Grades K–2* · ISBN: 978-1-55379-797-5

Portage & Main Press, 2019 · Hands-On Science for British Columbia · Land, Water, and Sky for Grades K–2 · ISBN: 978-1-55379-797-5

5

Explore Part Two

Conduct a book walk with *A Canadian Year: Twelve Months in the Life of a Canadian Kid*, by Tania McCartney (see lesson 2 for information on how to conduct a book walk).

Have students share their own stories of family and cultural celebrations during the different months of the year.

Explore Part Three

NOTE: The following activity will run over a period of one month and is an effective daily calendar routine to do with students. Use this opportunity to incorporate technology into your daily routine by creating a calendar on an interactive whiteboard or other device (e.g., computer, tablet). Calendar routines are also available from Promethean Planet <https://prometheanplanet.com/> or the SMART Exchange <exchange.smarttech.com/>.

Show students the large calendar for the current month. Ask:

- What month are we in now?
- On what date does the month begin?

Model the proper way to read the date (e.g., Today is Monday, October fourth). Have students repeat the date every day. Also, have students state the dates related to yesterday and tomorrow.

Focus on the day's weather. Ask:

- What is the weather like today?
- Is it sunny or cloudy?
- Is it snowing or raining?
- Is it windy or calm?

Print, cut out, and display the Picture Cards: Weather and Special Events (4.5.1):

Weather and Special Events

Weather and Special Events (continued)

Hands-On Science for British Columbia

5

Download this template at <www.portageandmainpress.com/product/HOSLandWaterSkyK2>.

Have students find a weather card that fits the day. Next, focus on upcoming events for the month. Ask:

- Does anyone have a birthday this month?
- Are there any special occasions this month?
- Do we have any field trips or special events planned for this month?

Add birthday and special-event cards to the calendar for the current month. Continue to focus on upcoming events for the month. Ask:

- How many days until (student's name) birthday?
- How many days are left in the month?

At the end of the month, review the weather, birthdays, and special events that occurred during the month to reinforce the concept of events that occur in a one-month period.

Use the monthly calendar to reinforce math concepts (e.g., counting, skip counting, before and after, patterns). Include these types of activities in your calendar study on a regular basis.

Expand

Provide students with an opportunity to explore the similarities and differences between the months further by posing their own questions for individualized inquiry. They may wish to:

- Initiate a project at the Makerspace, such as designing and constructing a desktop calendar.
- Explore Loose Parts bins related to the months with collections of objects that reflect the different months of the year (students may also participate in collecting these objects). Have students use the Loose Parts

to demonstrate what they know, can do, and/or understand about months of the year.

- Research and examine calendars from other cultures (e.g., the 13 Moons calendar, the Chinese calendar, the Islamic calendar). Students may wish to focus on their own cultural origin and research a specific calendar.
- Collect a variety of calendar pictures to examine, discuss, sequence, and sort.
- Design and construct a calendar. Illustrate pictures to depict each month of the year.
- Collect calendar stickers (calendars often come with them; ask families for donations). Students could make up an activity schedule for themselves, or a make-believe one on blank calendar pages.
- Research special celebrations around the world in a particular month (see page 26 for more information about inquiry through research).
- Make a picture book showing activities students do during each month. Have students present their books during sharing time.
- Have students who participate in particular cultural celebrations, share customs and traditions with the class when it occurs.
- Conduct an investigation based on their own inquiry questions.

As students explore and select ideas to expand learning, provide support and guidance as needed, and offer access to materials and resources that will enable students to conduct their chosen investigations.

Learning Centre

NOTE: Before students work at this learning centre, read *Chicken Soup With Rice: A Book of Months* by Maurice Sendak to focus further on the months of the year.

▶

Portage & Main Press, 2019 · *Hands-On Science for British Columbia · Land, Water, and Sky for Grades K–2* · ISBN: 978-1-55379-797-5

Portage & Main Press, 2019 · Hands-On Science for British Columbia · Land, Water, and Sky for Grades K–2 · ISBN: 978-1-55379-797-5

5

At the learning centre, provide a copy of *Chicken Soup With Rice: A Book of Months*, sample calendars, recycled scrap paper, and a copy of Learning-Centre Task Card: Acting Out the Months (4.5.2):

Learning Centre

Acting Out the Months

The months of the year go in a sequence, and each month is a little different from the other months.

1. Look through the pages of the book *Chicken Soup With Rice*, and read the poem for each month.

2. Choose a poem that you would like to act out.

3. By yourself or in a group, create actions to go with the words of the poem.

4. Share your actions with the rest of the class.

4.5.2

Download this template at <www.portageandmainpress.com/product/HOSLandWaterSkyK2>.

Have students select a month and create actions to go with their selected month. Have students present their actions to each other.

Embed Part One: Sharing Circle

Revisit the guided inquiry question: **How are the months of the year the same and different?** Have students share their experiences and knowledge, provide examples, and ask further inquiry questions.

Embed Part Two

- Add to the concept web as students learn new concepts, answer some of their own inquiry questions, and ask new inquiry questions.

- Add new terms and illustrations to the class word wall. Include the words in languages other than English, as appropriate.

- Focus on students' use of the Core Competencies. Have students reflect on how they used one of the Core Competencies (Thinking, Communicating, or Personal and Social Skills) during the various lesson activities. Project one of the CORE COMPETENCY DISCUSSION PROMPTS templates (pages 38–42), and use it to inspire group reflection. Referring to the template, choose one or two "I Can" statements on which to focus. Students then use the "I Can" statements to provide evidence of how they demonstrated that competency. Ask questions directly related to that competency to inspire discussion. For example:

  - How did you grow as a learner today? (Positive Personal and Cultural Identity)

  Have students reflect orally, encouraging participation, questions, and the sharing of evidence. (See page 29 for more information on these templates.) As part of this process, students can also set goals. For example, ask:

  - What would you do differently next time and why?

  - How will you know if you are successful in meeting your goal?

- To encourage self-reflection, provide prompts that students can use to cite examples of how they have used the Core Competencies in their learning. For this purpose, the CORE COMPETENCY SELF-REFLECTION FRAMES (pages 43–47) can be used throughout the learning process. There are five frames provided to address the Core Competencies: Communication, Creative Thinking, Critical Thinking, Positive Personal and Cultural Identity, and Personal Awareness and Responsibility. Teachers can conference individually with students to support self-reflection, or students may complete prompts using words and pictures. Again, have students set goals by considering what they might do differently on future tasks and how they will know if they are successful in meeting their goal.

NOTE: Use the same prompts from these templates over time to see how thinking changes with different activities.

Enhance

- **Family Connections:** Provide students with the following sentence starter:
  - Special family months are _____ because _____.

  Have students complete the sentence starter at home. Family members can help students draw and write about this topic. Have students share their sentences with the class.
- Make a large pictograph that identifies the birth month of each student in the class. As a class, select one of the following that students will draw to represent themselves on the pictograph: happy face, portrait, cupcake, birthday-cake slice (see pictograph example on page 24), or another picture the class chooses. When the pictograph has been completed, have students discuss and compare the data and record questions about the data for their classmates to answer. For example:
  - How many students have a birthday in June?
  - Which month has the most birthdays?
  - Which month has the least? The same?

- Over a period of one month, record the daily weather. Then, as a class, construct a pictograph of the month's weather. Use the weather cards from Weather and Special Events (4.5.1) for the pictograph.

Portage & Main Press, 2019 · Hands-On Science for British Columbia · Land, Water, and Sky for Grades K–2 · ISBN: 978-1-55379-797-5

6 | How Do We Know the Sun Gives Us Heat?

Information for Teachers

The Sun is the closest star to Earth; it keeps Earth warm. The Sun's rays are called *solar radiation*, and they are made of heat and light energy. The rays shine on Earth, travelling through the atmosphere, warming the air and Earth's land and oceans. Some radiation is reflected off Earth, back into space. Some energy is trapped in Earth's atmosphere, keeping the Earth warm.

 SAFETY NOTE: Instruct students to never look directly at the Sun, because it can damage their eyes.

Materials

- chart paper
- markers
- computer/tablet with internet access (optional)
- *How Raven Stole the Sun* by Maria Williams
- ice, ice cream, or something that will melt in a short period of time
- two bowls, labelled *A* and *B* (to hold the ice or ice cream)
- weather thermometers (one for each working group)
- outdoor thermometer
- digital camera (optional)
- concept web (from lesson 3)

 SAFETY NOTE: Use red alcohol thermometers rather than silver mercury thermometers. Mercury thermometers are no longer permitted in schools due to the toxic nature of this element.

Engage

Engage students in a discussion about the Sun. Ask:

- Where is the Sun?
- Can you touch the Sun?
- What does the Sun look like from where you are?
- What are the characteristics of the Sun?
- How does the Sun help us?

Title a sheet of chart paper "What We Know About the Sun." Record students' ideas about how the Sun looks, how it feels, and what the Sun does. Their ideas might include:

- The Sun is in the sky.
- The Sun looks round.
- The Sun is hot.
- The Sun gives us light.
- The Sun keeps us warm.
- You can get a sunburn if you are in the Sun for too long.

Encourage students to use vocabulary related to the Sun (e.g., *ray, sunlight, sunshine, warm, hot, daytime, nighttime, sunrise, sunset*).

Introduce the guided inquiry question: **How do we know the Sun gives us heat?**

Explore Part One

Read *How Raven Stole the Sun* by Maria Williams. Raven is often depicted as a trickster or troublemaker in Indigenous stories, but the results of the trickster's actions are often beneficial. In this story, the Raven releases the Sun, bringing light and heat to the world.

After reading the story, have a class discussion that connects the story to what students already know about the Sun. Ask:

- Why was it so important for Raven to release the Sun?
- How did the Earth and people stay warm before the Sun?
- Would plants be able to grow without the Sun?
- Would you survive without the Sun?

Portage & Main Press, 2019 · *Hands-On Science for British Columbia · Land, Water, and Sky for Grades K–2* · ISBN: 978-1-55379-797-5

There are many versions of the raven story that students can compare and contrast. For example:

- "Raven Steals the Light" <https://www.youtube.com/watch?v=74Y38Oy4AM4>
- *Strong Stories Coast Salish: The Sun and the Moon* by Celestine Aleck

Explore Part Two

Review with students what they already know about the Sun. Have students meet with a partner and share things they learned from previous lessons about the Sun (e.g., the relationship between the Sun and weather). Teachers may consider conducting a Think, Pair, Share activity to review prior knowledge.

NOTE: Students use thermometers for this activity. It is not necessary for students to actually read the temperature in degrees Celsius. However, they can begin to understand how thermometers work and how they show a rise and fall in temperature. Refer to the temperature as being where the red line is, and explain that the higher the red line goes, the higher/hotter the temperature.

Organize the class into working groups. Give each group a thermometer to examine. Have one student in each group be the "thermometer holder." Ask students to hold their thermometers near the middle of the tube, and to not touch the tip. Ask:

- What are these called?
- For what do we use them?
- How do we use them?

Have students examine the liquid inside the thermometer and look at the calibrated numbers along the side of the glass tube. Encourage them to recognize that a thermometer uses numbers, just like a number line. On a thermometer, however, the number line is vertical. This explanation may help students better understand the concept of how a thermometer works.

Compare the various thermometers to determine if the red liquid is at the same point (number) on each thermometer (the temperature on each should be close if students have not been touching the tips of the thermometers). Now, tell students to hold the tips of their thermometers and observe what happens.

Explore Part Three

Tell students they are going to do an experiment to learn more about the Sun and to find out if it gives us heat.

Place ice cream or ice into two bowls, labelled *A* and *B*. Place bowl A in direct sunlight, close to the window; place bowl B in a dark location of the classroom. Ensure students can see both bowls. Have students place a thermometer inside each bowl. Ask:

- What do you think will happen to the thermometer in each bowl?

Record students' predictions on chart paper. Test students' predictions: leave the bowls for several hours. Without touching or removing the thermometers, have students observe the bowls at regular intervals (e.g., every half hour). At each interval, ask:

- What has happened to the thermometers?
- Are both thermometers showing the same temperature?
- Which thermometer shows the higher temperature?
- Which thermometer shows the lower temperature?
- How do your observations compare with your predictions?

▶

Portage & Main Press, 2019 · *Hands-On Science for British Columbia · Land, Water, and Sky for Grades K–2* · ISBN: 978-1-55379-797-5

Portage & Main Press, 2019 · Hands-On Science for British Columbia · Land, Water, and Sky for Grades K–2 · ISBN: 978-1-55379-797-5

6

NOTE: If available, use interval-timed photography to create a time lapse of this experiment. Tablets have several apps to choose from: <http://appadvice.com/appguides/show/time-lapse-photography-apps>.

Have students draw a labelled diagram of the experiment.

As a follow-up, place an outdoor thermometer outside a classroom window. Students can observe temperature changes over one day and from day to day. Track the daily temperature as part of your monthly calendar routine.

Explore Part Four

Review the results of the experiment. Ask:

- What did the experiment teach you about the Sun?
- What special role does the Sun have for all living things?

Have students share their ideas and growing knowledge. Record their ideas on chart paper.

Formative Assessment

Observe students as they conduct the experiment. Focus on students' ability to discuss observations and draw diagrams of the experiment. Also focus on students' ability to describe how the Sun provides us with heat. Use the ANECDOTAL RECORD template, on page 50, to record results. Provide descriptive feedback to students about how they shared their observations through diagrams.

Expand

Provide students with an opportunity to explore the Sun further by posing their own questions for individualized inquiry. They may wish to:

- Initiate a project at the Makerspace (see pages 19 and 61), such as designing and constructing a hat to provide protection for skin and eyes.

- Explore Loose Parts bins related to the Sun. Students might use Loose Parts to show something they learned from Raven's story.
- Make a collection of objects and products related to the Sun (e.g., sunglasses with tinted lenses of different colours, visors, folding hand fans, empty sunscreen containers) to examine and discuss.
- Collect photographs of sunrises and sunsets, then paint either one with watercolours.
- Research the solar eclipse and construct a model (see page 26 for more information about inquiry through research).
- Learn about why it is important to wear sunglasses.
- Find out why we should never look directly at the Sun.
- Research the Sun.
- Learn about how sunscreen works to protect our skin.
- Conduct an investigation based on their own inquiry questions.

As students explore and select ideas to expand learning, provide support and guidance as needed, and offer access to materials and resources that will enable students to conduct their chosen investigations.

Embed Part One: Sharing Circle

Revisit the guided inquiry question: **How do we know the Sun gives us heat?** Have students share their knowledge and experience, provide examples, and ask further inquiry questions.

Embed Part Two

- Add to the concept web as students learn new concepts, answer some of their own inquiry questions, and ask new inquiry questions.
- Add new terms and illustrations to the class word wall. Include the words in languages other than English, as appropriate.

6

- Focus on students' use of the Core Competencies. Have students reflect on how they used one of the Core Competencies (Thinking, Communicating, or Personal and Social Skills) during the various lesson activities. Project one of the CORE COMPETENCY DISCUSSION PROMPTS templates (pages 38–42), and use it to inspire group reflection. Referring to the template, choose one or two "I Can" statements on which to focus. Students then use the "I Can" statements to provide evidence of how they demonstrated that competency. Ask questions directly related to that competency to inspire discussion. For example:

  What are you proud of in your learning today? (Personal Awareness and Responsibility) Have students reflect orally, encouraging participation, questions, and the sharing of evidence. (See page 29 for more information on these templates.)

  As part of this process, students can also set goals. For example, ask:

  - What would you do differently next time and why?
  - How will you know if you are successful in meeting your goal?

- To encourage self-reflection, provide prompts that students can use to cite examples of how they have used the Core Competencies in their learning. For this purpose, the CORE COMPETENCY SELF-REFLECTION FRAMES (pages 43–47) can be used throughout the learning process. There are five frames provided to address the Core Competencies: Communication, Creative Thinking, Critical Thinking, Positive Personal and Cultural Identity, and Personal Awareness and Responsibility. Teachers can conference individually with students to support self-reflection, or students may complete prompts using words and pictures. Again, have students set goals by considering what

they might do differently on future tasks and how they will know if they are successful in meeting their goal.

NOTE: Use the same prompts from these templates over time to see how thinking changes with different activities.

Enhance

- **Family Connections:** Provide students with the following sentence starter:
  - We keep track of the temperature at home by _____.

  Have students complete the sentence starter at home. Family members can help students draw and write about this topic. Have students share their sentences with the class.

- Introduce the concept of UV radiation and the need for Sun safety. Use UV sensing bracelets or beads if available. UV beads can also be used to test different sunscreen to see how they block UV radiation. Go to: <www.stevespanglerscience.com/store/uv-color-changing-beads.html>.

Portage & Main Press, 2019 · Hands-On Science for British Columbia · Land, Water, and Sky for Grades K–2 · ISBN: 978-1-55379-797-5

7 How Does the Temperature Change Throughout the Day?

Materials

- outdoor thermometers
- chart paper
- markers
- Template: Paper Thermometers (4.7.1)
- Template: Thermometer Timeline (4.7.2)
- Learning-Centre Task Card: A Temperature Timeline! (4.7.3)
- crayons
- glue
- scissors
- concept map (from lesson 3)

 SAFETY NOTE: Use red alcohol thermometers rather than silver mercury thermometers. Mercury thermometers are no longer permitted in schools due to the toxic nature of this element.

Engage

Record the term *temperature* on chart paper. Ask:

- What do you know about temperature?
- How is temperature measured?
- What did you learn about temperature when we did the experiment about the Sun's heat?
- What do you know about temperature and the seasons of the year?
- Do you think the temperature changes at different times of the day?

Record students' ideas on chart paper.

Introduce the guided inquiry question: **How does the temperature change throughout the day?**

Explore Part One

Explain to students that they will be measuring the outdoor temperature in the school yard at different times throughout the day for a five-day school week. As a class, design a recording sheet to track this data. For example:

Day of the Week	Temperature in the School Yard (degrees Celsius)			
	Morning	Lunch	After School	Night
Monday				
Tuesday				
Wednesday				
Thursday				
Friday				

Have students measure the temperature from the same location each time and record results, with guidance and assistance, as required.

NOTE: Ask for a student volunteer to record the temperatures in the evening at home.

When the data has been collected, review the results. Ask:

- Did you think the temperature would change or not change at different times during the day?
- Was your prediction correct?
- What time of the day was the hottest? Why do you think that was?
- What time of the day was the coldest?
- Why do you think that was?
- What would be different if you did this experiment during a different season?

Have students share their ideas and record them on chart paper.

Portage & Main Press, 2019 · Hands-On Science for British Columbia · Land, Water, and Sky for Grades K–2 · ISBN: 978-1-55379-797-5

Explore Part Two

Repeat this investigation at different times during the school year to investigate how temperature fluctuations are the same/different as seasons change.

Student Self-Assessment

Have students reflect on their investigation into temperature and complete journal entries to communicate their learning by recording observations. Have students use the SCIENCE JOURNAL template on page 37.

Expand

Provide students with an opportunity to explore daily temperature changes further by posing their own questions for individualized inquiry. They may wish to:

- Initiate a project at the Makerspace, such as designing and constructing a container to keep drinking water cool.
- Research the history of the thermometer (see page 26 for more information about inquiry through research).
- Examine different types of thermometers and research their uses.
- Research and experiment to find out when the hottest time of day is in a particular month. Students may be surprised to find out that the hottest time of the day is often during the late afternoon.

> ⚠️ **SAFETY NOTE:** Although the hottest time of the day is often later in the afternoon, remind students that doctors recommend staying out of the Sun between 10:00 AM and 2:00 AM when the Sun is directly overhead. Doctors also recommend using protective measures such as sunscreen, a sun hat, and sunglasses when going outside.

- Create a picture book about Sun safety.
- Use the data from the Explore experiments to construct a pictograph or bar graph.
- Create a picture book about temperature changes and related activities.
- Conduct an investigation or experiment based on their own inquiry questions.

As students explore and select ideas to expand learning, provide support and guidance as needed, and offer access to materials and resources that will enable students to conduct their chosen investigations.

Learning Centre

At the learning centre, include scissors, crayons, glue, copies of the Template: Paper Thermometers (4.7.1), and copies of the Template: Thermometer Timeline (4.7.2), along with a copy of the Learning-Centre Task Card: A Temperature Timeline! (4.7.3):

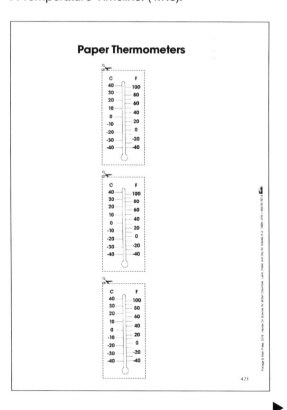

Portage & Main Press, 2019 · *Hands-On Science for British Columbia · Land, Water, and Sky for Grades K–2* · ISBN: 978-1-55379-797-5

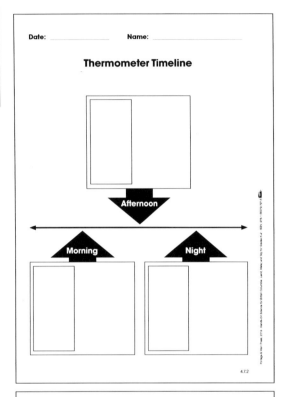

Download these templates at
<www.portageandmainpress.com/product/
HOSLandWaterSkyK2>.

Have students create timelines of the day, which include temperatures associated with three different times of day.

NOTE: Ensure an outdoor thermometer is accessible to students to read the current temperature. Also, check with students about whether or not they have an outdoor thermometer at home so they can read the temperature in the evening. Discuss alternatives, such as accessing the temperature online or from the Weather Network.

Embed Part One: Talking Circle

Revisit the guided inquiry question: **How does the temperature change throughout the day?** Have students share their experience and knowledge, provide examples, and ask further inquiry questions.

Embed Part Two

- Add to the concept web as students learn new concepts, answer some of their own inquiry questions, and ask new inquiry questions.
- Add new terms and illustrations to the class word wall. Include the words in languages other than English, as appropriate.
- Focus on students' use of the Core Competencies. Have students reflect on how they used one of the Core Competencies (Thinking, Communicating, or Personal and Social Skills) during the various lesson activities. Project one of the CORE COMPETENCY DISCUSSION PROMPTS templates (pages 38–42), and use it to inspire group reflection. Referring to the template, choose one or two "I Can" statements on which to focus. Students then use the "I Can" statements to provide evidence of how

Portage & Main Press, 2019 · Hands-On Science for British Columbia · Land, Water, and Sky for Grades K–2 · ISBN: 978-1-55379-797-5

they demonstrated that competency. Ask questions directly related to that competency to inspire discussion. For example:

- How did you make safe choices today? (Personal Awareness and Responsibility)

Have students reflect orally, encouraging participation, questions, and the sharing of evidence. (See page 29 for more information on these templates.) As part of this process, students can also set goals. For example, ask:

- What would you do differently next time and why?
- How will you know if you are successful in meeting your goal?

- To encourage self-reflection, provide prompts that students can use to cite examples of how they have used the Core Competencies in their learning. For this purpose, the CORE COMPETENCY SELF-REFLECTION FRAMES (pages 43–47) can be used throughout the learning process. There are five frames provided to address the Core Competencies: Communication, Creative Thinking, Critical Thinking, Positive Personal and Cultural Identity, and Personal Awareness and Responsibility. Teachers can conference individually with students to support self-reflection, or students may complete prompts using words and pictures. Again, have students set goals by considering what they might do differently on future tasks and how they will know if they are successful in meeting their goal.

NOTE: Use the same prompts from these templates over time to see how thinking changes with different activities.

Enhance

- **Family Connections**: Provide students with the following sentence starter:
  - We know the temperature changes throughout the day because _____.

  Have students complete the sentence starter at home. Family members can help students draw and write about this topic. Have students share their sentences with the class.
- Have students collect data from online weather sites and compare the data to classroom data.

Portage & Main Press, 2019 · *Hands-On Science for British Columbia · Land, Water, and Sky for Grades K–2* · ISBN: 978-1-55379-797-5

8 | What Do We Know About Seasonal Changes?

Information for Teachers

Climate in British Columbia varies greatly, and typical seasonal changes are not consistent throughout the province. Climate in British Columbia is influenced by latitude, mountainous topography, and the Pacific Ocean. This causes wide ranges in average rainfall, snowfall, temperature, and hours of sunshine. Therefore a wide range in seasonal weather patterns can be found in British Columbia.

Winters along the coast are temperate, and if snow falls, it does not stay long. Most of British Columbia's interior, on the other hand, experiences freezing temperatures and snow lasting from November to March.

Daytime temperatures in spring and fall can often be very warm and pleasant, especially June and September—particularly in southwestern British Columbia and the southern interior.

Summers are hottest in British Columbia's interior, particularly in the south where temperatures are often above 30°C. Nearer to the coast, temperatures range from 22–28°C.

For more information about British Columbia's climate, go to: <https://www.hellobc.com/british-columbia/about-bc/climate-weather.aspx>.

NOTE: This lesson will need to be adapted according to the local climate. For example, seasonal changes may be dramatic, affecting human activity as well as plants and animals. In other areas of the provinces, these seasonal changes may be subtler and may not affect living things as drastically.

Materials

- Image Bank: Seasons (print several copies of each image)
- place-based journals and supplies (see page 61 for more information)
- portable whiteboard (or chart paper with sturdy backboard)
- Indigenous stories about the creation of seasons*
- computer with disk drive or a DVD player (optional)
- computer/tablet with internet access
- printer (and paper)
- digital cameras
- concept web (from lesson 3)

*NOTE: The following DVDs are excellent resources: *Stories from the Seventh Fire: Summer and Autumn* and *Stories from the Seventh Fire: Winter and Spring* and can be ordered at <www.greenplanetfilms.org/product/stories-of-the-seventh-fire-summer-autumn/> and <www.greenplanetfilms.org/product/stories-of-the-seventh-fire-winter-spring/>. Another interesting site is Athabasca Chipewyan First Nation (Dene Suline/Soline) Literature. Go to <www.indigenouspeople.net/chipewyn.htm>.)

Engage

Organize the class into pairs and provide each pair with one printed copy of one image from the Image Bank: Seasons. Have pairs examine and discuss their image, and then present it to the class.

Display all of the printed images. Ask:

- What is different between these pictures?
- What is the weather like in each season?
- How are these seasons similar?
- How are these seasons different?
- How would you sort these pictures by season?
- How could you put these pictures in order?

Discuss and explore how the pictures can be sorted in different ways, and how they may be sequenced to show the order of the seasons.

Portage & Main Press, 2019 · Hands-On Science for British Columbia · Land, Water, and Sky for Grades K–2 · ISBN: 978-1-55379-797-5

8

NOTE: When events happen in a cycle (e.g., the seasons), students may recognize there is more than one possible starting point to the sequence.

Introduce the guided inquiry question: **What do we know about seasonal changes?**

Explore Part One

Read stories from local Indigenous communities about the creation of the seasons to students. Invite a local Elder or Knowledge Keeper to share stories.

NOTE: See Indigenous Perspectives and Knowledge, page 9, for guidelines for inviting Elders and Knowledge Keepers to speak to students.

Next, if possible, watch the following films:

- *Stories from the Seventh Fire: Summer and Autumn*
- *Stories from the Seventh Fire: Winter and Spring*

NOTE: This series, based on art by Norval Morrisseau and the live action footage of wildlife cinematographer Albert Karvonen, outlines Cree stories set in four seasons. The series was produced in English and Cree and uses 2D and 3D animation and the voices of some of Canada's best-loved Indigenous performers (e.g., Tantoo Cardinal, Gordon Tootoosis).

Also explore stories of Indigenous peoples in various parts of Canada such as the Dene. See "Dene: Creation of the Seasons" at: <www.indigenouspeople.net/chipewyn.htm>.

Explore Part Two

NOTE: If possible, conduct this activity in the fall, and then repeat it throughout the year, so students can observe changes through each season.

Plan a nature walk to observe the features of the current season. Consider inviting a local Elder or Knowledge Keeper to guide the walk and to share knowledge. Before the walk, be sure to review with students the importance of being respectful of nature when collecting objects. For example, branches should not be broken off trees. Small objects (e.g., twigs, leaves, seeds) may be taken in limited amounts and only with permission. Review the anchor chart created in lesson 1.

As with all place-based learning activities:

- Identify the importance of place. Use a map of the local area to identify where the location is in relation to the school.
- Identify on whose Indigenous territory the school is located, as well as the Indigenous territory of the location for the nature walk, if different. (See lesson 1).
- Incorporate land acknowledgment using local protocols (See lesson 1).

During the walk, have students observe the weather, plants, and animals. Have them photograph their observations.

To prepare students for journaling, complete a journal entry together as a class. This will require a portable whiteboard, or chart paper with a sturdy backboard for writing.

As a class, choose a place to sit and journal. Brainstorm journaling ideas. At the kindergarten to grade-two level, journaling can involve many different activities. Students may want to:

- sketch or colour the land, water, and sky they see around them
- sketch flowers or leaves
- identify emotions and record feelings about the current season (students may use happy faces, emojis, or their own designs for emotions)
- record the sounds they hear, using pictures and words
- record the movements they see, using pictures and words

▶

Portage & Main Press, 2019 · *Hands-On Science for British Columbia · Land, Water, and Sky for Grades K–2* · ISBN: 978-1-55379-797-5

Portage & Main Press, 2019 · Hands-On Science for British Columbia · Land, Water, and Sky for Grades K–2 · ISBN: 978-1-55379-797-5

8

- take a photograph that shows characteristics of the season and sketch it
- draw labelled diagrams showing features of the season
- use all their senses to describe the season using pictures or words
- write a poem about what they see or feel about the season they are experiencing

After working through a few journaling activities together, distribute place-based journals and supplies to students, have them choose a place where they would like to sit and journal, and have them choose one journaling strategy to record.

Explore Part Three

Back in the classroom, print off students' photographs from the nature walk. As a class, discuss the photographs. Ask:

- What was the weather like on our walk?
- What did you observe about animals?
- What did you observe about plants?
- What season is it now? How do you know?

Discuss students' background knowledge of the seasons. On chart paper, construct a table as in the following example:

Spring	Summer
Fall	Winter

Record students' ideas about the seasons.

Repeat this nature walk during each season to observe changes in weather, plants, animals, and human activity. Have students choose one thing to photograph (e.g., tree, shrub, creek, nest) during each season to make observations and draw comparisons.

Expand

Provide students with an opportunity to explore seasonal changes further by posing their own questions for individualized inquiry. They may wish to:

- Initiate a project at the Makerspace, such as designing and constructing a model of a favourite outdoor space during the four seasons.
- Create a graphic organizer to sort and display images of the seasons. This may include various templates used throughout this module and other classroom activities (e.g., Venn diagrams; See, Think, Wonder; KWHL charts; concept maps/webs).
- Create a picture book about the seasons or how to dress in each season.
- Draw their favourite outdoor space changing in each season and show a different activity that can be done there in each season.
- Conduct an investigation or experiment based on their own inquiry questions.
- Use place-based journaling to record ideas about seasonal changes (see lesson 1 for more information on place-based journals).

As students explore and select ideas to expand learning, provide support and guidance as needed, and offer access to materials and resources that will enable students to conduct their chosen investigations.

Embed Part One: Talking Circle

Revisit the guided inquiry question: **What do we know about seasonal changes?** Have students share their experience and knowledge, provide examples, and ask further inquiry questions.

Embed Part Two

- Add to the concept web as students learn new concepts, answer some of their own inquiry questions, and ask new inquiry questions.

- Add new terms and illustrations to the class word wall. Include the words in languages other than English, as appropriate.

- Focus on students' use of the Core Competencies. Have students reflect on how they used one of the Core Competencies (Thinking, Communicating, or Personal and Social Skills) during the various lesson activities. Project one of the CORE COMPETENCY DISCUSSION PROMPTS templates (pages 38–42), and use it to inspire group reflection. Referring to the template, choose one or two "I Can" statements on which to focus. Students then use the "I Can" statements to provide evidence of how they demonstrated that competency. Ask questions directly related to that competency to inspire discussion. For example:

  - How did you show that you were an active listener today? (Communication)

  Have students reflect orally, encouraging participation, questions, and the sharing of evidence. (See page 29 for more information on these templates.) As part of this process, students can also set goals. For example, ask:

  - What would you do differently next time and why?

  - How will you know if you are successful in meeting your goal?

- To encourage self-reflection, provide prompts that students can use to cite examples of how they have used the Core Competencies in their learning. For this purpose, the CORE COMPETENCY SELF-REFLECTION FRAMES (pages 43–47) can be used throughout the learning process. There are five frames

provided to address the Core Competencies: Communication, Creative Thinking, Critical Thinking, Positive Personal and Cultural Identity, and Personal Awareness and Responsibility. Teachers can conference individually with students to support self-reflection, or students may complete prompts using words and pictures. Again, have students set goals by considering what they might do differently on future tasks and how they will know if they are successful in meeting their goal.

NOTE: Use the same prompts from these templates over time to see how thinking changes with different activities.

Enhance

- **Family Connections**: Provide students with the following sentence starters:

  - Some of the seasonal changes my family enjoys are _____ because_____.

  - Some of the activities my family enjoys in each of the seasons are _____.

  Have students complete the sentence starter at home. Family members can help students draw and write about this topic. Have students share their sentences with the class.

- Have students search online for pictures of each season. Have them print the pictures and add them to the word wall.

Portage & Main Press, 2019 · Hands-On Science for British Columbia · Land, Water, and Sky for Grades K–2 · ISBN: 978-1-55379-797-5

9 | How Do Seasonal Changes Affect Plants?

Information for Teachers

In some areas of Canada and British Columbia, deciduous, or broadleaf, trees lose their leaves in the fall when the amount of sunlight decreases and the days become cooler. Harsh winter conditions may bring frost, snow, and ice, and the ground becomes hard. Trees become bare, and nearby plants enter a dormant stage. Tree growth slows down during this time. In spring, trees begin to bud and new leaves appear. New grass and flowers around the trees begin to grow, and the soil and air temperatures feel warmer under the trees. Trees remain green for most of the summer, but the soil and plant life beneath them will start to dry.

NOTE: Animals that inhabit or frequent deciduous trees change throughout the seasons, as well.

Plants used by Indigenous peoples in British Columbia throughout the seasons include:

Food plants: Hundreds of species of wild plants can be gathered for food. These include seeds, nuts, and grains (e.g., hazelnuts, acorns, whitebark pine seeds, wild rice). Berries and fleshy fruits (e.g., blueberries, huckleberries, crabapples, rosehips, wild cherries) are important sources of vitamins. So are wild greens, the shoots and leaves of various plants, and leaf vegetables (e.g., mustard greens, lamb's quarters [pigweed], watercress). Roots are harvested later in the season (e.g., wild onion, balsam root, wild turnip). Fungi, such as wild mushrooms, and certain tree barks are also be eaten.

In the interior of British Columbia, berries, roots, mushrooms, lichen, and inner bark are important to First Peoples. Diet depends on location. There is a temperate rainforest on the coast and climate similar to desert in the interior.

Medicinal plants: Many plants can be used to treat illnesses and ailments. Some are administered as teas, such as Labrador tea, used for kidney ailments, or yarrow, to treat colds and fever. Some plants are used as inhalants or as poultices applied to a certain part of the body (e.g., a poultice from the purple coneflower root treats sores and swelling). Others are mixed with fats to make ointments. Devil's club, for example, has multiple uses (e.g., poultices and soaps, chewed for vitamins).

There are endless lists of medicinal plants and each is specific to a given community and the environment in which they live.

Sacred plants: Like the Anishinaabe in Ontario, the peoples in British Columbia value tobacco, sweetgrass, sage, and cedar, among other plants, as part of sacred ceremonies. The cedar is especially important for the Gitxsan, who associate purity and cleansing with cedar because of its rot-resistant properties. Gitxsan babies are born on a cedar mat, and in death, the Gitxsan are buried in a cedar box. The root of the Indian hellebore—called malgwasxw by the Gitxsan—is used in smudging ceremonies, where the smoke produced is part of prayer and cleansing ceremonies. It is believed that the smoke acts as a messenger to the spirit world. Sage, sweetgrass, tobacco, and cedar may also be used as offerings for prayer and ceremony.

Utility plants: Plants are essential for day-to-day life. Wood from various trees is used to build structures (e.g., longhouses) and is used for transportation (e.g., canoes, snowshoe frames, toboggans). Wood is also used for tools and implements (e.g., arrows and spear shafts, digging sticks, chisel and adze handles, toys and games). Bark is used to create containers and canoes, and can be used as wrapping and lining. Plant fibres are used to make rope, fabric, mats, and baskets. Moss can be used to line diapers. Many different plants are used as dyes.

Portage & Main Press, 2019 · Hands-On Science for British Columbia · Land, Water, and Sky for Grades K–2 · ISBN: 978-1-55379-797-5

9

As previously mentioned, cedar is very important for First Nations people in British Columbia, depending on location.

Materials

- *Four Seasons Make a Year* by Anne Rockwell (or another book that focuses on living things throughout the seasons)
- chart paper
- markers
- digital camera
- ribbon or string (optional)
- scissors (optional)
- Image Bank: Indigenous Plants Throughout the Seasons (see Appendix, page 165)
- books about plants and trees (see Explore Part Four)
- Learning-Centre Task Card: How Can We Model Trees Throughout the Seasons? (4.9.1)
- concept web (from lesson 3)

Engage

Read *Four Seasons Make a Year* by Anne Rockwell, or another book about living things and the seasons. Discuss the events and ideas in the book. Ask:

- What do you know about the seasons?
- What season is it now?
- What are the names of the four seasons?
- What is the order of the four seasons?
- What happens to plants in the spring?
- What happens to plants in the summer?
- What happens to plants in the fall?
- What happens to plants in the winter?
- Why do you think the seasons change?
- How does the weather change during the seasons?
- Why do you think the weather changes during the seasons?
- What is the temperature like in each of the seasons?

Have students share their responses.

Introduce the guided inquiry question: **How do seasonal changes affect plants?**

Explore Part One

Take a walk around the local area to examine plants.

Before the walk, be sure to review with students the importance of being respectful of nature when collecting objects. For example, branches should not be broken off trees. Small objects (e.g., twigs, leaves, seeds) may be taken in limited amounts and only with permission. Review the anchor chart created in lesson 1.

As with all place-based learning activities:

- Identify the importance of place. Use a map of the local area to identify the location in relation to the school.
- Identify on whose Indigenous territory the school is located, as well as the Indigenous territory of the location for the nature walk, if different (see lesson 1).
- Incorporate land acknowledgment using local protocols (see lesson 1).

During the walk, have students identify and describe the plants, and take digital pictures of the plants. During the walk, ask:

- What season is it now?
- Will this plant be affected as the seasons change? How?

During the walk, find a natural space to conduct a role-play activity. Have students use large and small muscle movement to dramatize how a tree would "act and feel" under the following conditions:

- gentle spring breeze
- violent autumn windstorm
- pelting rain
- summer forest fire

▶

Portage & Main Press, 2019 · *Hands-On Science for British Columbia · Land, Water, and Sky for Grades K–2* · ISBN: 978-1-55379-797-5

Portage & Main Press, 2019 · Hands-On Science for British Columbia · Land, Water, and Sky for Grades K–2 · ISBN: 978-1-55379-797-5

9

- having bare limbs in the winter
- squirrel running up its trunk
- bird nesting in its branches
- being climbed by a person
- being cut down by a person

After the role-playing activity, discuss the various ways trees change throughout the seasons.

Explore Part Two

NOTE: If possible, conduct this activity in the fall, and then repeat it throughout the year, so students can observe changes through each season.

Tell students that they will be going outside to examine a deciduous tree in the school yard or local community.

Before the nature walk, be sure to review with students the importance of being respectful of nature when collecting objects. Review the anchor chart created in lesson 1.

As with all place-based learning activities:

- Identify the importance of place. Use a map of the local area to identify the location in relation to the school.
- Identify on whose Indigenous territory the school is located, as well as the Indigenous territory of the location for the nature walk, if different (see lesson 1).
- Incorporate land acknowledgment using local protocols (see lesson 1).

At the tree, discuss the appearance of the tree, and photograph it. If possible, mark the tree with a ribbon so you can revisit the same tree throughout the year to photograph it as it changes through the seasons. Ask:

- What season is it now?
- What are the characteristics of trees in (name season)?
- What happens to trees during the different seasons?

- Do all trees change?
- Which trees do not seem to change from season to season?
- What could you watch for as you study the tree from season to season?
- How are the seasons like a pattern? (e.g., ABCD, spring, summer, fall, winter)
- How are the changes in some trees like a pattern? (e.g., buds, leaves, leaves turning and falling, no leaves)

Have students share their ideas. Discuss the differences between deciduous and coniferous trees, and highlight the variation in seasonal changes in different parts of British Columbia.

Explore Part Three

Organize the class into four working groups. Give each group a sheet of chart paper, and assign each group one of the four seasons. Ask the groups to develop a poster about their assigned season. Work with students to co-construct criteria for the posters. For example:

- includes the name of the season
- includes a large labelled diagram of the tree (as a class, decide which parts are to be labelled [e.g., leaves, branches, trunk, bark, roots])
- describes and illustrates the weather of the season
- explains how the season affects this tree

Have students work together in their groups to create posters that depict characteristics of their assigned season—with the tree as the focal point of the image.

Have each group of students present their poster to the class, and display the posters throughout the school year. As each season arrives, have students compare their posters to the actual season, the weather, and their tree.

9

Explore Part Four

Explore local Indigenous peoples' uses of plants throughout the seasons. Invite a local Elder or Knowledge Keeper to share stories and information about this topic, such as:

- how to pull cedar bark
- how to harvest sage
- seasonal rounds
- seeding in spring and harvesting in fall
- picking blueberries in the spring and summer
- harvesting maple sap in the spring
- peeling birch bark from the trees to make baskets, houses, and canoes in the summer

NOTE: See Indigenous Perspectives and Knowledge, page 9, for guidelines for inviting Elders and Knowledge Keepers to speak to students.

Display the Image Bank: Indigenous Plants Throughout the Seasons. Have students examine each picture, describe the plant's characteristics, and determine which season it might be. Use Information for Teachers to discuss how these plants are used by Indigenous peoples in British Columbia. Also, read books such as:

- *Kawlija's Blueberry Promise* by Audrey Guiboche tells of the summer blueberry harvest of a Métis family.
- *A Journey Through the Circle of Life* by Desiree Gillespie celebrates the planting of trees each year to honour Mother Nature.
- *Wild Berries* by Julie Flett portrays the process of blueberry picking.
- *The Apple Tree* by Sandy Tharp-Thee tells the story of a contemporary Cherokee boy who plants an apple seed and already sees the mature apple tree it is meant to be.
- *Niwechihaw: I Help* by Caitlin Dale Nicholson is a story in Cree and English that explores a young child's relationship to his kuhkom

(grandmother), as they go for a walk in the bush to pick rosehips in the fall.

Learning Centre

At the learning centre, provide pictures of deciduous trees in all four seasons and digital cameras (and video cameras if students are able to use them, or if an older peer/adult is able to record the role-play activities). Also include a copy of the Learning-Centre Task Card: How Can We Model Trees Throughout the Seasons? (4.9.1):

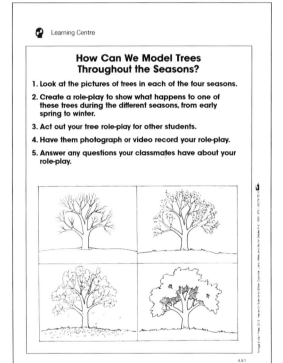

Download this template at <www.portageandmainpress.com/product/HOSLandWaterSkyK2>.

Have students role-play what happens to a deciduous tree during the different seasons, from early spring to winter. Teachers may first want to model a few other examples (e.g., role-

▶

Portage & Main Press, 2019 · *Hands-On Science for British Columbia · Land, Water, and Sky for Grades K–2* · ISBN: 978-1-55379-797-5

9

play a bird from egg to chick to bird in flight; a flower from seed to seedling to a flower in bloom). Students can then create their own role-plays to depict a tree during the different seasons.

Expand

Provide students with an opportunity to explore seasonal effects on local plants further by posing their own questions for individualized inquiry. They may wish to:

- Initiate a project at the Makerspace, such as designing and constructing models of a tree or other plant in the four seasons.
- Explore Loose Parts related to seasonal changes in plants in the local area (e.g., acorns, leaves, twigs, bark, seeds, seedpods). Students can explore the Loose Parts to generate their own inquiry questions and to personalize learning experiences.
- Use place-based journaling to express and reflect on personal experiences of place with regard to the plants observed on the nature walk.
- Create their own graphic organizer to sort and display pictures of local plants. This may include various templates used throughout this module and other classroom activities (e.g., Venn diagrams; See, Think, Wonder; KWHL chart; concept maps/webs).
- Explore watercolour painting to depict trees/plants in various seasons.
- Create a picture book about local plants and how they change throughout the seasons.
- Conduct an investigation or experiment based on their own inquiry questions.

As students explore and select ideas to expand learning, provide support and guidance as needed, and offer access to materials and resources that will enable students to conduct their chosen investigations.

Formative Assessment

Focus on students making simple predictions about seasonal effects on plants. During individual interviews, ask students to identify the seasons in order and to give words that remind them of each season. Present them with a variety of images of plants during different seasons, and have students make predictions about how these plants will change throughout the seasons. Record results for each student on the INDIVIDUAL STUDENT OBSERVATIONS template on page 51. Provide descriptive feedback to students about how they identified, described, and made predictions about the seasons.

Student Self-Assessment (PC)

Have students take home a copy of the FAMILY AND COMMUNITY CONNECTIONS ASSESSMENT template on page 57. Have them complete the template with a family or community member (with permission) to reflect on their learning about how seasonal changes affect plants, by recording observations.

Embed Part One: Sharing Circle

Revisit the guided inquiry question: **How do seasonal changes affect plants?** Have students share their experience and knowledge, provide examples, and ask further inquiry questions.

Embed Part Two

- Add to the concept web as students learn new concepts, answer some of their own inquiry questions, and ask new inquiry questions.
- Add new terms and illustrations to the class word wall. Include the words in languages other than English, as appropriate.
- Focus on students' use of the Core Competencies. Have students reflect on how they used one of the Core Competencies (Thinking, Communicating, or Personal

Portage & Main Press, 2019 · Hands-On Science for British Columbia · Land, Water, and Sky for Grades K–2 · ISBN: 978-1-55379-797-5

9

and Social Skills) during the various lesson activities. Project one of the CORE COMPETENCY DISCUSSION PROMPTS templates (pages 38–42), and use it to inspire group reflection. Referring to the template, choose one or two "I Can" statements on which to focus. Students then use the "I Can" statements to provide evidence of how they demonstrated that competency. Ask questions directly related to that competency to inspire discussion. For example:

- What are your strengths in learning about plants and seasonal changes? (Positive Personal and Cultural Identity)

Have students reflect orally, encouraging participation, questions, and the sharing of evidence. (See page 29 for more information on these templates.) As part of this process, students can also set goals. For example, ask:

- What would you do differently next time and why?
- How will you know if you are successful in meeting your goal?

- To encourage self-reflection, provide prompts that students can use to cite examples of how they have used the Core Competencies in their learning. For this purpose, the CORE COMPETENCY SELF-REFLECTION FRAMES (pages 43–47) can be used throughout the learning process. There are five frames provided to address the Core Competencies: Communication, Creative Thinking, Critical Thinking, Positive Personal and Cultural Identity, and Personal Awareness and Responsibility. Teachers can conference individually with students to support self-reflection, or students may complete prompts using words and pictures.

Again, have students set goals by considering what they might do differently on future tasks and how they will know if they are successful in meeting their goal.

NOTE: Use the same prompts from these templates over time to see how thinking changes with different activities.

Enhance

- **Family Connections**: Provide students with the following sentence starter:
  - A plant that we have observed that changed during the seasons is _____.

  Have students complete the sentence starter at home. Family members can help students draw and write about this topic. Have students share their sentences with the class.

- Teach students the words for the four seasons in a local Indigenous language to reinforce and validate the importance of Indigenous languages. For example:

English	Northern St'át'imcets
Spring	Qapts
Summer	pipántsek
Fall	lhwáltsten
Winter	Sútik

NOTE: The spelling and pronunciation of words can vary between communities, depending on the dialect of the language. Consider contacting local Indigenous language organizations for accuracy of spelling and pronunciation or invite an Elder or Knowledge Keeper experienced with a local language into the classroom.

- Use seasonal vocabulary from languages other than English reflective of the classroom community. Include these on the word wall.

- Using students' photographs of trees, have students write Compare/Contrast notes about any two photographs, and choose their favourite.

- If students are from other parts of the world, have them compare British Columbia's seasonal changes to those from their place of origin (e.g., Bengali kids will be familiar with the rainy season. When is it? How long

▶

Portage & Main Press, 2019 · Hands-On Science for British Columbia · Land, Water, and Sky for Grades K–2 · ISBN: 978-1-55379-797-5

9

does it last? What is occurring in our British Columbia community at the same time?).

- Throughout the school year, read books, learn songs and chants, and discuss the seasonal changes that occur in trees, animals, and the weather.

- Have students play the game Four Corners, using words for the seasons. Have students brainstorm a list of words for each season (e.g., fall: cool, windy, golden, crisp; winter: snow, ice, cold, snowman). In each corner of the classroom or gym, place a card with the name of one of the four seasons (include Indigenous words for the seasons, as well). Call out a word from one of the season's lists. Students must move to the corner of the room labelled with the season to which the word relates.

- Make a class Big Book for each of the four seasons to place in the classroom or school library. Provide students with large sheets of paper, along with pencils and crayons. Have each student create one page for each book by writing and drawing depictions of the weather and the behaviour of living things during that particular season.

- Record, on index cards, the words that students brainstormed for each of the seasons (also consider including pictures on these cards to support students' reading). Have students sort the words into the four seasons. Display the index cards in a pocket chart and make title cards for each grouping.

- Investigate how some plants cannot survive drastic changes in temperature, and discuss how humans help to protect some plants from seasonal extremes (e.g., wrapping cedars in burlap to prevent wind damage, wrapping wire around shrubs so that deer cannot eat them, digging up tubers and bulbs that would freeze in the ground over the winter).

Portage & Main Press, 2019 · Hands-On Science for British Columbia · Land, Water, and Sky for Grades K–2 · ISBN: 978-1-55379-797-5

10 | How Do Seasonal Changes Affect Animals?

Materials

- Image Bank: Animals of British Columbia (see Appendix, page 165)
- resources about animals during the different seasons (e.g., picture books, videos, nonfiction books)
- computers/tablets with internet access
- projection device
- chart paper
- markers
- art paper
- sheets of cardboard (approx. 60 cm², divided into quarters with a black marker) (one for each working group)
- play dough or Plasticine
- materials for making habitat models (e.g., sand, soil, twigs, rocks, grass, blue cellophane for water, cotton batting for snow, egg cartons for hills/mountains)
- Learning-Centre Task Card: Some Final Thoughts About My Project (4.10.1)
- audio-recording device
- concept web (from lesson 3)

Engage

Display the Image Bank: Animals of British Columbia. As each image is projected, have students identify the animal and discuss its characteristics. Based on students' growing background knowledge, have them discuss the role these animals have in life in British Columbia and how the animals are important to Indigenous peoples (e.g., food, hides, feathers, fur, bones, teeth).

For each animal, identify the season in which the photograph was taken. Review with students the four seasons of the year and characteristics of each season. Also, have students share background knowledge related to how the animal responds to seasonal changes.

Introduce the guided inquiry question: **How do seasonal changes affect animals?**

Explore Part One

Read various books with students about changes in animals throughout the four seasons. Then, title four sheets of chart paper with the headings "Animals in Spring," "Animals in Summer," "Animals in Fall," and "Animals in Winter." Record students' ideas about animal behaviours and characteristics during each of the seasons on the appropriate sheet. Display the sheets in a location for all students to see.

Explore Part Two

Organize the class into working groups, and have each group select an animal to study. Their task will be to design a model depicting their animal during the various seasons. For example, if a group is studying geese, students may decide to show the geese flying north in the spring, living in a marsh in the summer, flying south in the fall, and living by a lake in the winter. Remind students that some animals hibernate.

When all groups have selected an animal to research, as a class, identify criteria for the inquiry. Criteria can be posed in the form of questions. Brainstorm questions based on ideas from the word wall and concept web. For example:

- What does our animal do in each season?
- What does our animal do to protect itself during the changing seasons?
- How does our animal's natural environment change with the seasons?
- A question of our own choice.

Model the process to scaffold learning (I Do, We Do, You Do). Select an animal (one not selected by students). Walk through the process of researching the animal to determine how it behaves during the various seasons. Model

Portage & Main Press, 2019 · *Hands-On Science for British Columbia · Land, Water, and Sky for Grades K–2* · ISBN: 978-1-55379-797-5

Portage & Main Press, 2019 · Hands-On Science for British Columbia · Land, Water, and Sky for Grades K–2 · ISBN: 978-1-55379-797-5

for students how to look for information in a nonfiction text. Use chart paper to record research notes in the form of text and labelled diagrams.

Next, have students follow this process. Provide a variety of sources and provide time for groups to research their selected animals and to record results.

Explore Part Three

When groups have completed their research, discuss the next activity—the creation of animal models. Explain that their job is to make models that show what their animal does in each season. They will first use Plasticine to create four miniature figurines of their animal. Show students one of the cardboard sheets. Explain that in each section of the cardboard, they will create a background to show what this animal might do in each of the four seasons.

As a class, co-construct criteria for the construction of the models. For example:

- Create a background to show characteristics of the four seasons.
- Show what the animal does in each season.
- Show how the animal protects itself in each season.
- Show one more interesting fact about the animal.

Model for students how to design and construct a sample project using a different animal than they have suggested. Have students make suggestions for creating the background and for meeting the criteria.

Next, distribute one cardboard sheet to each group, along with Plasticine and materials for making habitat models (e.g., sand, soil, twigs, rocks, grass, blue cellophane for water, cotton batting for snow, egg cartons for hills/ mountains). Give groups time to complete

their models. When completed, have each group present their animal models to the class. Together, review project criteria to determine if their projects meet the criteria.

Expand

Provide students with an opportunity to explore how seasonal changes affect animals further by posing their own questions for individualized inquiry. They may wish to:

- Initiate a project at the Makerspace, such as designing and constructing a model of a zoo enclosure that depicts seasonal changes.
- Explore Loose Parts related to how seasonal changes affect animals. Have students represent their animal during different seasons, or have them explore other animal habitats in various seasons using Loose Parts (e.g., plastic animals, rocks, acorns, pine cones, sticks, leaves, soil). Loose Parts may also be used on the sand or water table for this exploration.
- Create their own graphic organizer to sort and display photographs of how seasonal changes affect local animals (e.g., hibernation, migration). Graphic organizers may include various templates used throughout this module and other classroom activities (e.g., Venn diagrams; See, Think, Wonder; KWHL; concept maps/webs).
- Create a picture book about local animals throughout the seasons.
- Make a puppet of a local animal throughout the seasons, and create a puppet show.
- Conduct an investigation or experiment based on their own inquiry questions.

As students explore and select ideas to expand learning, provide support and guidance as needed, and offer access to materials and resources that will enable students to conduct their chosen investigations.

▶

10

Learning Centre

At the learning centre, provide an audio-recording device, along with a copy of each group's animal habitat project, and a copy of the Learning-Centre Task Card: Some Final Thoughts About My Project (4.10.1):

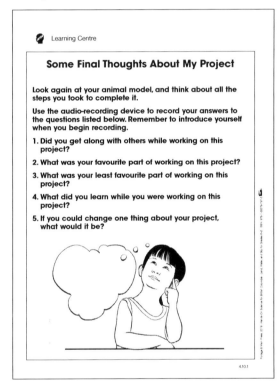

Download this template at <www.portageandmainpress.com/product/HOSLandWaterSkyK2>.

Have students audio record their self-reflections based on an assessment of their own projects.

NOTE: To expand this activity, use a screen casting app, or use Pages or Keynote to create a digital book with a voice recording.

Formative Assessment (AI)

Record criteria for the inquiry projects on the RUBRIC template, on page 53, and record results for each student. Meet with students individually to have them share how they met the criteria for

the project. Focus on students' ability to transfer and apply learning to new situations. Provide descriptive feedback to students about how they applied learning and met the project criteria.

Embed Part One: Sharing Circle

Revisit the guided inquiry question: **How do seasonal changes affect animals?** Have students share their experiences and knowledge, provide examples, and ask further inquiry questions.

Embed Part Two

- Add to the concept web as students learn new concepts, answer some of their own inquiry questions, and ask new inquiry questions.
- Add new terms and illustrations to the class word wall. Include the words in languages other than English, as appropriate.
- Focus on students' use of the Core Competencies. Have students reflect on how they used one of the Core Competencies (Thinking, Communicating, or Personal and Social Skills) during the various lesson activities. Project one of the CORE COMPETENCY DISCUSSION PROMPTS templates (pages 38–42), and use it to inspire group reflection. Referring to the template, choose one or two "I Can" statements on which to focus. Students then use the "I Can" statements to provide evidence of how they demonstrated that competency. Ask questions directly related to that competency to inspire discussion. For example:
  - How did you get new ideas about how animals are affected by seasonal changes? (Critical Thinking)

Have students reflect orally, encouraging participation, questions, and the sharing of evidence. (See page 29 for more information on these templates.)

Land, Water, and Sky for Grades K–2   **117**

Portage & Main Press, 2019 · *Hands-On Science for British Columbia · Land, Water, and Sky for Grades K–2* · ISBN: 978-1-55379-797-5

As part of this process, students can also set goals. For example, ask:

- What would you do differently next time and why?
- How will you know if you are successful in meeting your goal?

- To encourage self-reflection, provide prompts that students can use to cite examples of how they have used the Core Competencies in their learning. For this purpose, the CORE COMPETENCY SELF-REFLECTION FRAMES (pages 43–47) can be used throughout the learning process. There are five frames provided to address the Core Competencies: Communication, Creative Thinking, Critical Thinking, Positive Personal and Cultural Identity, and Personal Awareness and Responsibility. Teachers can conference individually with students to support self-reflection, or students may complete prompts using words and pictures. Again, have students set goals by considering what they might do differently on future tasks and how they will know if they are successful in meeting their goal.

NOTE: Use the same prompts from these templates over time to see how thinking changes with different activities.

Enhance

- **Family Connections**: Provide students with the following Sentence starter:
  - One animal we see affected by changes in the seasons is _____.

  Have students complete the sentence starter at home. Family members can help students draw and write about this topic. Have students share their sentences with the class.

- Have students use Shadow Puppet or 30 Hands to capture the models of their animals during the four seasons. Students can then create a screen capture with voice-over to represent their knowledge and understanding.

11 | Which Activities Do People Do During Different Seasons?

Information for Teachers

This lesson, in part, focuses on Indigenous peoples' knowledge and use of seasonal rounds. According to the British Columbia Science Curriculum (2017), "seasonal rounds refers to a pattern of movement from one resource-gathering area to another in a cycle that is followed each year."

NOTE: A valuable resource on this topic is "My Seasonal Round." Go to <https://www.openschool.bc.ca/elementary/my_seasonal_round/pdf/SeasonalRound_unit.pdf>.

Materials

- Song Lyrics: "Seasons of the Year" (Print, project, or copy onto chart paper.) (4.11.1)
- projection device (optional)
- chart paper
- markers
- resources about the seasonal impact on life of Indigenous peoples in British Columbia. (See Explore Part One)
- computer/tablet with internet access
- drawing paper
- art supplies (e.g., pencil crayons, crayons, paints, paintbrushes)
- clipboards
- rulers
- Learning-Centre Task Card: What Are Our Favourite Seasonal Activities? (4.11.2)
- Learning-Centre Survey: Which Is Your Favourite Seasonal Activity? (4.11.3)
- concept web (from lesson 3)

Engage

Project (or display on chart paper) the Song Lyrics: "Seasons of the Year" (4.11.1):

Seasons of the Year
by Meish Goldish

(sing to the tune of "Here We Go 'Round the Mulberry Bush")

CHORUS:
Here we go round the year again,
The year again, the year again.
Here we go round the year again,
To greet the different seasons.

Wintertime is time for snow.
To the south, the birds will go.
It's too cold for plants to grow
Because it is the winter.

CHORUS

In the springtime, days grow warm.
On the plants, the new buds form.
Bees and bugs come out to swarm
Because it is the spring.

CHORUS

In summertime, the days are hot.
Ice cold drinks I drink a lot!
At the beach, I've got a spot
Because it is the summer.

CHORUS

Fall is here, the air is cool.
Days are short, it's back to school.
Raking leaves is now the rule
Because it is autumn.

CHORUS

From *101 Science Poems and Songs for Young Readers* by Meish Goldish. Copyright © 1996 by Meish Goldish. Reprinted by permission of Scholastic Inc.

4.11.1

Download this template at <www.portageandmainpress.com/product/HOSLandWaterSkyK2>.

Teach the lyrics to students. Now, ask students:

- What is the weather like in spring/summer/fall/winter?
- What do humans do when it is windy/rainy/snowy/hot/cold?
- What do we wear during different seasons?
- What activities do humans do in different seasons?
- How do humans prepare for the next season? (e.g., get out seasonal clothing, prepare the yard, tune up snow blower or lawn mower, find rake or shovel)
- Who lets us know about what the temperature will be?
- Where can we find information about the daily weather?

▶

Portage & Main Press, 2019 · *Hands-On Science for British Columbia · Land, Water, and Sky for Grades K–2* · ISBN: 978-1-55379-797-5

Portage & Main Press, 2019 · Hands-On Science for British Columbia · Land, Water, and Sky for Grades K–2 · ISBN: 978-1-55379-797-5

11

Introduce the guided inquiry question: **Which activities do people do during different seasons?**

Explore Part One

As a class, learn about seasonal rounds. Consider inviting a local Elder or Knowledge Keeper to share information and stories related to seasonal rounds.

NOTE: See Indigenous Perspectives and Knowledge, page 9, for guidelines for inviting Elders and Knowledge Keepers to speak to students.

As a class, watch the following videos:

- "Winter Homes" <https://www.youtube.com/watch?v=bzJ-mLcWiXs>
- "Summer Home Materials" <https://www.youtube.com/watch?v=H1GwODsJgSg>

Also explore resources such as:

- "Resource Gathering" <http://secwepemc.sd73.bc.ca/sec_village/sec_round.html>
- "My Seasonal Round" <https://www.openschool.bc.ca/elementary/my_seasonal_round/pdf/SeasonalRound_unit.pdf>

Explore stories of Indigenous peoples and how the seasons influence daily life. For example:

- *Neekna and Chemai* by Jeanette Armstrong. Two girls grow up in the Okanagan area of the province. Each chapter talks about traditional practices they participate in during the various seasons.
- *Byron Through the Seasons: A Dene-English Story Book*, by the students of La Loche. This book follows a grandfather's stories of his traditional activities during the seasons.

- *Things That Keep Us Warm* by Louise Flaherty. The author describes how people stay warm during winter. The book highlights items common throughout Canada, such as the parka, and some items that are uniquely northern, such as the Inuit qulliq (oil lamp).
- *Did You Hear the Wind Sing Your Name?: An Oneida Song of Spring* by Sandra de Coteau Orie. This book explains Oneida worldviews, including the importance of the Hawk, the bringer of good news; the sustaining Elder Brother, the Sun; the use of cedar and sweetgrass in ceremonies; and the Three Sisters: corn, beans, and squash.
- *Skysisters* by Jan Bourdeau Waboose. On a cold winter's night in Northern Ontario, two Ojibway sisters set out in search of the Sky Spirits their mother has told them will come that night to do their sky dance.

Read the stories to the class, and examine and discuss the illustrations that correspond to each season. Provide students with drawing paper and art supplies, and have them create their own illustrations to correspond with a selected story.

Explore Part Two

Title four sheets of chart paper with the headings "Humans in Spring," "Humans in Summer," "Humans in Fall," and "Humans in Winter." Discuss, and then record, activities that students and their families participate in during each of the four seasons. Discuss students' favourite seasonal activities, and include these ideas on the charts.

NOTE: Be sure to include activities that are done during all seasons, such as walking or hiking.

Focus on how humans can do certain activities out of season. Ask:

- What would you do if you wanted to go swimming in the middle of winter?
- Would you go to the beach?

- Where could you go?
- What if you wanted to skate during the summer?
- Would there be ice outside?
- Where could you go?

To reinforce the various activities done during the seasons, play a version of Charades. Have a student select and act out an activity from one of the charts (e.g., downhill skiing, playing baseball, shovelling snow, raking leaves). Challenge the rest of the students to guess the activity being acted out and identify the season.

Expand

Provide students with an opportunity to explore human activities during the seasons further by posing their own questions for individualized inquiry. They may wish to:

- Initiate a project at the Makerspace, such as designing and constructing a model of the equipment needed for a specific seasonal activity (e.g. toboggan, downhill skis and poles, motorboat and water-skis, snowshoes).
- Explore Loose Parts bins related to seasonal activities. Have students use Loose Parts to show what they have learned about seasonal rounds.
- Make collections of objects related to a specific season (e.g., sidewalk chalk of different colours, bubble wands, bike reflectors, spoke decorations).
- Research a specific hobby or seasonal activity, and create a poster or slideshow (see page 26 for more information about inquiry through research).
- Make puppets, and then create a puppet show of favourite seasonal activities.

- Explore various sand/beach toys (e.g., shovels, pails, sand moulds) at the sand table.
- Explore various toy boats and/or toy fishing rods at the water table.
- Conduct an investigation or experiment based on personal inquiry.

As students explore and select ideas to expand learning, provide support and guidance as needed, and offer access to materials and resources that will enable students to conduct their chosen investigations.

Learning Centre

At the learning centre, provide rulers, clipboards, a copy of Learning-Centre Task Card: What Are Our Favourite Seasonal Activities? (4.11.2), and copies of the Learning-Centre Survey: Which Is Your Favourite Seasonal Activity? (4.11.3):

Portage & Main Press, 2019 · Hands-On Science for British Columbia · Land, Water, and Sky for Grades K–2 · ISBN: 978-1-55379-797-5

Portage & Main Press, 2019 · Hands-On Science for British Columbia · Land, Water, and Sky for Grades K–2 · ISBN: 978-1-55379-797-5

11

Date: _____   **Name:** _____

Which Is Your Favourite Seasonal Activity?

Season: _____

Activity	Tally	Total

4.11.3

Download these templates at <www.portageandmainpress.com/product/HOSLandWaterSkyK2>.

Have students survey their classmates to determine their favourite activities during one of the seasons. Tell students to use words, pictures, charts, or graphs to show their results. For graphing, students may use the tally data to construct a concrete-objects graph for one-to-one correspondence or a pictograph.

Embed Part One: Sharing Circle

Revisit the guided inquiry question: **Which activities do people do during different seasons?** Have students share their experience and knowledge, provide examples, and ask further inquiry questions.

Embed Part Two

- Add to concept web as students learn new concepts, answer some of their own inquiry questions, and ask new inquiry questions.
- Add new terms and illustrations to the class word wall. Include the words in languages other than English, as appropriate.
- Focus on students' use of the Core Competencies. Have students reflect on how they used one of the Core Competencies (Thinking, Communicating, or Personal and Social Skills) during the various lesson activities. Project one of the CORE COMPETENCY DISCUSSION PROMPTS templates (pages 38–42), and use it to inspire group reflection. Referring to the template, choose one or two "I Can" statements on which to focus. Students then use the "I Can" statements to provide evidence of how they demonstrated that competency. Ask questions directly related to that competency to inspire discussion. For example:
  - How did you share your learning with others today? (Communication)
  Have students reflect orally, encouraging participation, questions, and the sharing of evidence. (See page 29 for more information on these templates.)
  As part of this process, students can also set goals. For example, ask:
  - What would you do differently next time and why?
  - How will you know if you are successful in meeting your goal?
- To encourage self-reflection, provide prompts that students can use to cite examples of how they have used the Core Competencies in their learning. For this purpose, the CORE COMPETENCY SELF-REFLECTION FRAMES (pages 43–47) can be used throughout the learning process. There are five frames provided to address the Core Competencies:

11

Communication, Creative Thinking, Critical Thinking, Positive Personal and Cultural Identity, and Personal Awareness and Responsibility. Teachers can conference individually with students to support self-reflection, or students may complete prompts using words and pictures. Again, have students set goals by considering what they might do differently on future tasks and how they will know if they are successful in meeting their goal.

NOTE: Use the same prompts from these templates over time to see how thinking changes with different activities.

Enhance

- **Family Connections**: Provide students with the following sentence starters:
    - My family prepares for winter/spring/summer/fall by _____.
    - Each spring/summer/fall/winter, my family _____.

    Have students complete the sentence starter at home. Family members can help students draw and write about this topic. Have students share their sentences with the class.

- Invite a local Elder or Knowledge Keeper to discuss with students how Indigenous peoples continue to take part in specific activities according to the seasons. Have students create posters showing activities that Indigenous peoples do during the different seasons. Distribute art paper and crayons. Students can draw a circle on the paper and then divide the circle into quadrants. Ask students to write the name of one season in each quadrant. Then have them write and glue or draw pictures of activities Indigenous peoples do in the different seasons (e.g., hunting in winter, collecting maple syrup in spring, planting gardens and picking raspberries in summer, harvesting wild rice and cranberries in fall).

NOTE: See Indigenous Perspectives and Knowledge, page 9, for guidelines for inviting Elders and Knowledge Keepers to speak to students.

- Discuss various weather extremes (e.g., blizzards, tornadoes, floods, heat waves, thunder-and-lightning storms). Have students review how to stay safe in these conditions and discuss who helps us prepare for these weather extremes (e.g., meteorologists, news reporters).

- Read and discuss books that focus on different activities in different seasons, from an Indigenous perspective. For example:
    - *Morning on the Lake* by Jan Bourdeau Waboose
    - *Lessons From Mother Earth* by Elaine McLeod

- Not all parts of the world experience the contrasting seasons that we experience in Canada. Have students research seasonal changes in other countries.

Portage & Main Press, 2019 · *Hands-On Science for British Columbia · Land, Water, and Sky for Grades K–2* · ISBN: 978-1-55379-797-5

12 | Which Objects Do We See in the Daytime Sky?

Information for Teachers

This lesson focuses on various patterns and cycles in the local sky, including the *Sun*, *rainbows,* and *clouds.*

Rainbows are created by refraction. When conditions are right, water droplets in the air can act to separate the colours in sunlight—the water droplets act like prisms, refracting and reflecting the sunlight and separating the colours. This creates the colours of the rainbow. Sometimes, rainbows are visible in a fine spray of water (e.g., from a lawn sprinkler). The water droplets act in the same manner as raindrops—refracting, reflecting, and dispersing light.

Clouds form above the surface of the Earth when air cools and water vapour condenses on dust particles in the air, forming tiny droplets of water. Millions of tiny droplets form a cloud (the cloud is made of water droplets or ice particles, depending on whether condensation occurred above or below the freezing point of water). The cloud grows bigger if more condensation than evaporation occurs. The cloud becomes smaller if more evaporation than condensation occurs.

There are three basic types of clouds: *cirrus*, *cumulus*, and *stratus*.

Stratus: sometimes look like a blanket stretched across the sky. They are usually located below 2500 metres. *Stratus* means *layer*. These clouds often bring rainy weather.

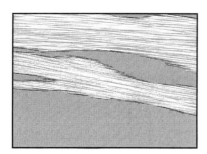

Cirrus: located the highest in the atmosphere (6000–18 000 metres above Earth). They are made of ice crystals. The word *cirrus* means *curl*. These wispy clouds look curly and travel very fast. They are often seen ahead of changes in the weather.

Cumulus: look puffy, like cotton balls with flat bottoms. They are found 1000–14 000 metres above Earth. *Cumulus* means *pile*. When the lower sections of these clouds begin to darken, a storm is often near.

Materials

- *Little Cloud* by Eric Carle
- *It Looked Like Spilt Milk* by Charles G. Shaw
- resources about clouds and rainbows
- place-based journals and supplies (see page 61 in for more information)
- Image Bank: Rainbows (see Appendix, page 165)
- portable whiteboard (or chart paper with sturdy backboard)
- Learning-Centre Task Card: The Nature of Clouds (4.12.1)

Portage & Main Press, 2019 · Hands-On Science for British Columbia · Land, Water, and Sky for Grades K–2 · ISBN: 978-1-55379-797-5

12

- white chalk, pencil crayons, or crayons
- scissors
- chart paper
- markers
- large sheets of blue construction paper
- glue
- cotton batting
- digital camera
- concept web (from lesson 3)

Engage

Have students lie down on the carpet and participate in a visual imagery activity. Recite the following (or use the text as a guide for your own personal interpretation):

- Close your eyes and let your body relax. Imagine that you are lying in the grass at your favourite place in nature, on a warm summer afternoon. What do you hear? What do you smell? What do you feel?

- Keep your eyes closed, but imagine that you are looking up at the sky in this special place. What might you see in the sky? Take a few minutes to imagine all of the things that you might see in the sky.

Have students sit up and share the experience with an elbow partner. Encourage them to discuss their special place in nature, as well as what they might hear, smell, feel, and see there.

As a class, discuss objects in the sky. Ask:

- What might you see in the sky in your special place?

Brainstorm and record students' ideas on chart paper.

Introduce the guided inquiry question: **Which objects do we see in the daytime sky?**

Explore Part One

Model a book walk for the class using *Little Cloud* by Eric Carle or *It Looked Like Spilt Milk* by Charles G. Shaw.

NOTE: See lesson 2 for more information about how to conduct a book walk.

Have students share their background knowledge of, and experiences with, clouds.

Explore Part Two

Explain to students that they will be going on a nature walk to explore objects in the sky. Discuss the location for the place-based learning. Ask:

- Who has been to this place before?
- How did you get there?
- What is it like there?
- What do you think we will see there? Smell? Hear? Feel?

Have students share their background knowledge, predictions, inferences, and ideas about the natural environment that they will visit. Record their ideas on chart paper to refer back to later to see how thinking can change and if they are able to find answers to their questions.

Have the students share with a classmate what they are most excited about in visiting this location for place-based learning.

Before the walk, be sure to review the importance of being respectful of nature when collecting objects. For example, branches should not be broken off trees. Small objects (e.g., twigs, leaves, seeds) may be taken in limited amounts and only with permission. Review the anchor chart created in lesson 1.

As with all place-based learning activities:

- Identify the importance of place. Use a map of the local area to identify where the location is in relation to the school.

Portage & Main Press, 2019 · *Hands-On Science for British Columbia · Land, Water, and Sky for Grades K–2* · ISBN: 978-1-55379-797-5

Portage & Main Press, 2019 · *Hands-On Science for British Columbia · Land, Water, and Sky for Grades K–2* · ISBN: 978-1-55379-797-5

- Identify on whose traditional territory the school is located, as well as the traditional territory of the location for the nature walk, if different (see lesson 1).
- Incorporate land acknowledgment using local protocols (see lesson 1).

Review any other protocols for field trips, providing students with opportunities to ask questions and clarify expectations.

⚠ SAFETY NOTE: Remind students not to look directly at the Sun when observing objects in the sky, as it could damage their eyes.

Explore Part Three

On the nature walk, provide an initial opportunity for students to explore the environment freely (with supervision). After, have them share their observations as to what they see, hear, feel, and smell.

⚠ SAFETY NOTE: Remind students never to taste objects without permission.

Next, challenge students to observe objects in the sky. Have them lie down and look at the sky (again, without looking directly at the Sun). Provide ample time for students to observe and reflect.

Next, have students work in groups to photograph what they observe in the sky (no photographing the Sun).

Meet as a group to discuss their findings.

Explore Part Four

Provide multiple opportunities for students to observe various types of clouds in the sky. Check weather reports and use observation to select days and times when students will see these clouds.

Have students take their place-based journals outdoors to sketch and write about different cloud formations. Also encourage them to photograph various cloud formations to use back in the classroom.

To prepare students for journaling, do a journal entry together as a class. This will require a portable white board and markers, or chart paper with a sturdy backboard.

As a class, choose a place to sit and journal. Brainstorm journaling ideas. At the kindergarten to grade-two level, journaling can involve many different activities. Students may want to:

- sketch or colour the clouds and sky
- identify emotions, recording feelings about the clouds and sky (students may use happy faces, emojis, or their own designs for emotions)
- record sounds they hear, using pictures and words
- record movements, using pictures and words
- photograph the clouds and sky and sketch them
- draw labelled diagrams showing the clouds and sky
- use all of their senses to describe the clouds and sky using pictures and/or words
- write a poem about what they observe or feel about the sky.
- draw and illustrate their own *Little Cloud* book based on the changes in shape they observe in the clouds

After working through a few journaling activities together, distribute place-based journals and supplies to students, have them choose a place where they would like to sit and journal, and have them choose one journaling strategy to record.

Following journaling activities in which students observe various cloud types, have them use related resources to compare their drawings and photographs to identify types of clouds.

Explore Part Five

Project the Image Bank: Rainbows. Have students examine and describe the various rainbows. Ask:

- When do we usually see rainbows?
- What do you think causes a rainbow?

Have students share background experiences and knowledge.

NOTE: This relates to the module, *Properties of Energy.*

Explore Indigenous stories related to rainbows. For example:

- *Rainbow Crow: English / Cree Edition* by David Bouchard
- *Enora and the Black Crane* by Arone Raymond Meeks

Have students use their growing knowledge to draw labelled diagrams of rainbows, showing their understanding of how and when rainbows occur.

Expand

Provide students with an opportunity to explore clouds and rainbows further by posing their own questions for individualized inquiry. They may wish to:

- Initiate a project at the Makerspace, such as designing and constructing a model to show how rainbows are formed. They might use lights, a light table, and various coloured objects and materials to create a rainbow.
- Explore Loose Parts related to colours of the rainbow with collections of objects that reflect the colours of the rainbow (e.g., marker caps, coloured gems, glass/wooden/plastic beads, mosaic tiles, crayons).

Challenge students to create rainbow patterns with the objects, and explore further based on their own inquiry questions and ideas. Also provide small mirrors on which students can build rainbows.

- Explore various art supplies related to the colours of the rainbow (e.g., water colours, tissue paper, crepe paper, Plasticine, pipe cleaners, fun foam).
- Create a graphic organizer to display diagrams and photographs of cloud types.
- Research how different clouds are formed, and create a presentation for the class (see page 26 for more information about inquiry through research).
- Research how rainbows are formed and create a presentation for the class.
- Research sundogs, how double rainbows occur, or which weather patterns create the different clouds.
- Use translucent materials of various colours (e.g. a light table, coloured beads, tinted Plexiglas) to explore the colours of the rainbow, and to recreate rainbows.
- Conduct an investigation or experiment based on personal inquiry.

As students explore and select ideas to expand learning, provide support and guidance as needed, and offer access to materials and resources that will enable students to conduct their chosen investigations.

Learning Centre

At the learning centre, provide large sheets of blue construction paper, glue, scissors, and cotton batting, along with their place-based journals, photographs, and resource materials related to types of clouds. Also provide white pencils crayons, white crayons, or chalk and a copy of the Learning-Centre Task Card: The Nature of Clouds (4.12.1):

▶

Portage & Main Press, 2019 · *Hands-On Science for British Columbia · Land, Water, and Sky for Grades K–2* · ISBN: 978-1-55379-797-5

Portage & Main Press, 2019 · *Hands-On Science for British Columbia · Land, Water, and Sky for Grades K–2* · ISBN: 978-1-55379-797-5

Download this template at <www.portageandmainpress.com/product/HOSLandWaterSkyK2>.

Have students use cotton batting to create various types of clouds. Have them use white crayons, pencil crayons or chalk to label or add other details to their cloud collages.

Embed Part One: Sharing Circle

Revisit the guided inquiry question: **Which objects do we see in the daytime sky?** Have students share their experiences and knowledge, provide examples, and ask further inquiry questions.

Embed Part Two

- Add to concept web as students learn new concepts, answer some of their own inquiry questions, and ask new inquiry questions.

- Add new terms and illustrations to the class word wall. Include the words in languages other than English, as appropriate.

- Focus on students' use of the Core Competencies. Have students reflect on how they used one of the Core Competencies (Thinking, Communicating, or Personal and Social Skills) during the various lesson activities. Project one of the CORE COMPETENCY DISCUSSION PROMPTS templates (pages 38–42), and use it to inspire group reflection. Referring to the template, choose one or two "I Can" statements on which to focus. Students then use the "I Can" statements to provide evidence of how they demonstrated that competency. Ask questions directly related to that competency to inspire discussion. For example:

  - How did you get new ideas as you learned today? (Creative Thinking)

  Have students reflect orally, encouraging participation, questions, and the sharing of evidence. (See page 29 for more information on these templates.)

  As part of this process, students can also set goals. For example, ask:

  - What would you do differently next time and why?

  - How will you know if you are successful in meeting your goal?

- To encourage self-reflection, provide prompts that students can use to cite examples of how they have used the Core Competencies in their learning. For this purpose, the CORE COMPETENCY SELF-REFLECTION FRAMES (pages 43–47) can be used throughout the learning process. There are five frames provided to address the Core Competencies: Communication, Creative Thinking, Critical Thinking, Positive Personal and Cultural Identity, and Personal Awareness and Responsibility. Teachers can conference individually with students to support self-reflection, or students may complete prompts using words and pictures. Again, have students set goals by considering what they might do differently on future tasks and how they will know if they are successful in meeting their goal.

NOTE: Use the same prompts from these templates over time to see how thinking changes with different activities.

Enhance

- **Family Connections**: Provide students with the following sentence starters:
    - My family has seen different clouds when _____.
    - My family has seen a rainbow when _____.

    Have students complete the sentence starters at home. Family members can help students draw and write about this topic. Have students share their sentences with the class.
- On a cold day, take students outside and have them create their own clouds. They can do this by blowing warm air from their lungs into the cold outside air. Have students describe what happens when they blow into the cold air. Ask:
    - How this like the formation of a cloud? (Water droplets are created when the warm vapour from the lungs meets the cold air outside.)
- Watch videos about cloud formation, such as "Luke Howard – the man who named the clouds" go to <https://www.youtube.com/watch?v=iWjBjE4lys8>.
- Read *The Man Who Named the Clouds* by Julie Hanna, to have the students learn more about the classifications of clouds and how they affect the weather.

- Discuss safety with regard to clouds students see in the sky. Ask:
    - When should you get out of the pool or off the soccer pitch?
- Use online drawing programs to create cloud pictures.

Portage & Main Press, 2019 · *Hands-On Science for British Columbia · Land, Water, and Sky for Grades K–2* · ISBN: 978-1-55379-797-5

---

13 | Which Objects Do We See in the Nighttime Sky?

Information for Teachers

The Sun and the objects that orbit the sun comprise our *solar system*. The largest objects orbiting the Sun are the eight planets: Mercury, Venus, Earth, Mars, Jupiter, Saturn, Uranus, and Neptune. The solar system also includes objects called dwarf planets, including Pluto, which until 2006 was considered the ninth planet. Other objects in the solar system include the following:

Moons: natural satellites orbiting planets. Moons reflect light from the Sun. Our moon orbits Earth over a period of a month (actually, 27.3 days), and as it does, it appears to change shape (from our perspective on Earth). This illusion occurs because only the side of the moon that faces the Sun is lit up. As the moon orbits Earth, we see different amounts of this lit-up side. When the moon is between the Sun and the Earth, we cannot see the moon because the dark side of it is facing us.

Asteroids: rocky or metallic minor planets orbiting the Sun. Most asteroids are found in an asteroid belt between Mars and Jupiter.

Meteoroids: chunks of rock, ice, and metal drifting in space. Meteoroids are often broken-off pieces from asteroids or remnants of comets.

Meteors: pieces of meteoroids that reach Earth's atmosphere and burn up, creating brilliant streaks in the sky. They are also called "shooting stars."

Meteorites: lumps of rock or metal that have fallen to the Earth's surface. Meteorites are the result of meteors that have not completely burned up.

Comets: bright masses of ice, frozen gases, and dust particles that orbit the Sun. Comets have one or more long, gaseous tails that point away from the Sun and are often visible on Earth.

Stars: luminous objects in the night sky. On a clear night, the naked eye can see about 3000 stars. Ancient astronomers divided the sky into groups of stars called *constellations*— clusters of stars in the sky that seem to take on recognizable patterns. Many constellations are named after mythological characters and events. Greek and Roman mythology gave rise to some names, in other cultures, different names were applied to star clusters.

Aurorae: phenomena related to Earth's magnetic traits. Aurorae are made when energetic gas particles from the Sun create a solar wind. Aurorae are most likely to be found at the poles of the planet, due to the poles' strong magnetic force, which attracts the gas particles. When these particles collide with the gases in Earth's atmosphere (mainly oxygen and nitrogen), the aurora's colourful display of light occurs (e.g., northern lights/aurora borealis).

Materials

- chart paper
- markers
- black paper
- white chalk or pastels
- resources about the Moon, stars, planets, and Aurorae (see Resources for Students)
- Indigenous stories about the night sky (see Explore Part One)
- *The Way Back Home* by Oliver Jeffers, *Papa, Please Get the Moon for Me* by Eric Carle, or another story about the night sky
- Image Bank: The Moon (see Appendix, page 165)
- concept web (from lesson 3)

Engage

Have students lie down on the carpet and participate in a visual imagery activity. Recite the following (or use the text as a guide for your own personal interpretation):

Portage & Main Press, 2019 · *Hands-On Science for British Columbia · Land, Water, and Sky for Grades K–2* · ISBN: 978-1-55379-797-5

13

- Close your eyes and let your body relax. Imagine you are lying in the grass at your favourite place in nature. It is nighttime. What do you hear? What do you smell? What do you feel?

- Keep your eyes closed, but imagine that you are looking up at the dark night sky in this special place. What might you see in the sky? Take a few minutes to imagine all of the things that you might see in the sky.

Have students sit up and share the experience with an elbow partner. Encourage them to discuss their special place in nature, as well as what they might hear, smell, feel, and see there in the night sky.

As a class, discuss objects in the night sky. Ask:

- What might you see in the night sky in your special place?

Brainstorm and record students' ideas on chart paper.

Provide each student with black paper and white chalk or pastels. Have them draw images that they might see in the local night sky.

Introduce the guided inquiry question: **Which objects do we see in the nighttime sky?**

Explore Part One

Explore Indigenous stories related to the night sky. Invite a local Elder or Knowledge Keeper to share information and stories with the class.

NOTE: See Indigenous Perspectives and Knowledge, page 9, for guidelines for inviting Elders and Knowledge Keepers to speak to students.

Use related resources to explore Indigenous perspectives, knowledge, and stories about the night sky. For example:

- *Tipiskawi Kisik: Night Sky Star Stories* by Wilfred Buck.
- *Strong Stories Kanyen'keha:ka: Big Bear* by Michelle Corneau
- *Orphans in the Sky* by Jeanne Bushey

Explore Part Two

Read *The Way Back Home* by Oliver Jeffers, or *Papa, Please Get the Moon for Me* by Eric Carle. As the book is read, have students identify and describe objects in the night sky. Ask:

- What do you know about the moon?

Have students share their ideas and record these on chart paper.

Project the Image Bank: The Moon. Have students examine and describe each image, sharing their ideas and background knowledge. Encourage students to identify similarities and differences between the images. Ask:

- What do you think the surface of the moon is like?
- Would it be easy to walk or ride your bike on the moon?
- What shape is the moon?
- Do the photographs always show the moon as round?
- Have you seen the moon in a shape other than round?

Have students describe the differences they have observed in the moon's shape. Record students' descriptions on chart paper, and have them draw diagrams to support their observations.

▶

Portage & Main Press, 2019 · Hands-On Science for British Columbia · Land, Water, and Sky for Grades K–2 · ISBN: 978-1-55379-797-5

13

Explore Part Three

Review the books presented in the previous sections of this lesson. Review the illustrations and have students identify stars in the night sky. Encourage students to share their background knowledge of stars, constellations, and personal experiences stargazing.

Encourage family participation to provide students with an authentic personalized learning experience to observe the night sky. Send home a letter to families. Explain the content of the lesson and request participation in the learning by planning a family stargazing night. Provide students with the following sentence starter:

- When we gaze at the night sky, we see _____.

Have students complete the sentence starter at home. Family members can help students draw and write about this topic. Also provide students with dark blue or black paper and white chalk or pastels to draw labelled diagrams of their observations. Have students share their sentences and diagrams with the class.

Explore Part Four

As a class, learn about the aurorae. Many videos offer the opportunity to view the northern lights, or aurora borealis. For example:

- "Night of the Northern Lights" <https://www.youtube.com/watch?v=fVsONlc3OUY>
- "Stunning Aurora Borealis from Space" <https://www.youtube.com/watch?v=fVMgnmi2D1w>

Consider having a local Elder or Knowledge Keeper speak to the class to share their knowledge of how the northern lights play a role in Indigenous life.

NOTE: This is likely only relevant for Nations that are up north. In lower BC, northern lights are not often visible.

There are also many stories about the northern lights from different Indigenous peoples, which can be shared with the class.

As a class, read an Indigenous story about the aurora borealis:

- *Strong Readers Northern Series: Look Up at the Sky!* By Brenda Boreham
- *Warren Whistles at the Sky* by David Alexander Robertson
- *Sky Sisters* by Jan Bourdeau Waboose

Expand

Provide students with an opportunity to explore objects in the night sky further by posing their own questions for individualized inquiry. They may wish to:

- Initiate a project at the Makerspace, such as designing and constructing a model of the moon or solar system.
- Explore Loose Parts bins related to the night sky with collections of star shapes from various materials (e.g., fun foam, foil, commercial glow-in-the-dark stars/planets) for students to examine, discuss, sort, and to inspire further inquiry.
- Create a graphic organizer to display diagrams and photographs of objects in the night sky.
- Research one of the planets in the solar system (see page 26 for more information about inquiry through research).
- Create a picture book of the phases of the moon.
- Conduct an investigation or experiment based on personal inquiry.

As students explore and select ideas to expand learning, provide support and guidance as needed, and offer access to materials and resources that will enable students to conduct their chosen investigations.

Portage & Main Press, 2019 · *Hands-On Science for British Columbia · Land, Water, and Sky for Grades K–2* · ISBN: 978-1-55379-797-5

Embed Part One: Sharing Circle

Revisit the guided inquiry question: **Which objects do we see in the nighttime sky?** Have students share their experiences and knowledge, provide examples, and ask further inquiry questions.

Embed Part Two

- Add to concept web as students learn new concepts, answer some of their own inquiry questions, and ask new inquiry questions.

- Add new terms and illustrations to the class word wall. Include the words in languages other than English, as appropriate.

- Focus on students' use of the Core Competencies. Have students reflect on how they used one of the Core Competencies (Thinking, Communicating, or Personal and Social Skills) during the various lesson activities. Project one of the CORE COMPETENCY DISCUSSION PROMPTS templates (pages 38–42), and use it to inspire group reflection. Referring to the template, choose one or two "I Can" statements on which to focus. Students then use the "I Can" statements to provide evidence of how they demonstrated that competency. Ask questions directly related to that competency to inspire discussion. For example:

  - How did you show how you can work with others today? (Communication)

  Have students reflect orally, encouraging participation, questions, and the sharing of evidence. (See page 29 for more information on these templates.)

  As part of this process, students can also set goals. For example, ask:

  - What would you do differently next time and why?

  - How will you know if you are successful in meeting your goal?

- To encourage self-reflection, provide prompts that students can use to cite examples of how they have used the Core Competencies in their learning. For this purpose, the Core COMPETENCY SELF-REFLECTION FRAMES (pages 43–47) can be used throughout the learning process. There are five frames provided to address the Core Competencies: Communication, Creative Thinking, Critical Thinking, Positive Personal and Cultural Identity, and Personal Awareness and Responsibility. Teachers can conference individually with students to support self-reflection, or students may complete prompts using words and pictures. Again, have students set goals by considering what they might do differently on future tasks and how they will know if they are successful in meeting their goal.

NOTE: Use the same prompts from these templates over time to see how thinking changes with different activities.

Enhance

- **Family Connection**: Have students conduct research to find pictures of constellations. These can be printed and taken home. Provide students with the following sentence starter:

  - We can find objects in the night sky such as _____.

  Have students complete the sentence starter at home. Family members can help students draw and write about this topic. Have students share their sentences with the class.

- Read and discuss such books as, *The Big Dipper* by Franklyn Branley or *How to Catch a Star* by Oliver Jeffers.

- After learning about the stories and pictures behind the constellations, have students create their own constellations using miniature white marshmallows and

▶

Portage & Main Press, 2019 · *Hands-On Science for British Columbia · Land, Water, and Sky for Grades K–2* · ISBN: 978-1-55379-797-5

Portage & Main Press, 2019 · Hands-On Science for British Columbia · Land, Water, and Sky for Grades K–2 · ISBN: 978-1-55379-797-5

toothpicks glued onto black paper. Students choose a constellation, draw it on the black paper, and then construct the constellation using the marshmallows to represent the stars and the toothpicks as the imaginary connecting lines.

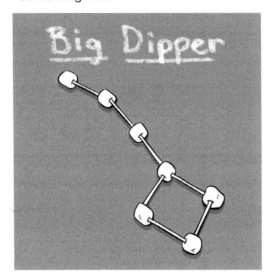

- Use SkyView, Star Chart or another app to view constellations through a tablet or other device.
- Visit a planetarium or observatory.

14 | What Forms of Water Are Found on Earth?

Materials

- world globe
- chart paper
- markers
- Indigenous stories about water (see Explore Part Two)
- digital camera
- wall map of British Columbia
- Image Bank: Forms of Water on Earth (print each image) (see Appendix, page 165)
- place-based journals and supplies (see page 61 for more information)
- concept web (from lesson 3)

Engage

Display the world globe. Ask students:

- What is this called?
- What does a globe show you?
- What do the green parts of the globe show?
- What do the blue parts of the globe show?
- Is there more water or land on the surface of Earth?
- What kinds of water, other than oceans, are there on Earth?
- Have you been to the ocean?
- What is the water like in the ocean?

Discuss the difference between saltwater in oceans and seas, and fresh water in lakes and rivers.

Introduce the guided inquiry question: **What forms of water are found on Earth?**

Explore Part One

Plan a nature walk to explore local bodies of water, such as a creek, pond, or stream.

 SAFETY NOTE: Ensure student safety around water at all times. Students should be closely supervised and should not be near water if there is a risk of slipping or falling in, especially larger bodies of water.

Small local creeks or ponds are more suitable for this exploration.

Before the walk, review with students the importance of being respectful of nature when collecting objects. Review the anchor chart created in lesson 1.

As with all place-based learning activities:

- Identify the importance of place. Use a map of the local area to identify where the location is in relation to the school.
- Identify on whose traditional territory the school is located, as well as the traditional territory of the location for the nature walk, if different (see lesson 1).
- Incorporate land acknowledgment using local protocols (see lesson 1).

Have students explore the features of this body of water. Have them generate inquiry questions before the water investigation. For example:

- What is the water's edge like?
- What kind of plants grow along the water?
- How clear is the water?
- What kinds of plants live in the water?
- What kinds of animals live in the water?

Distribute placed-based journals and supplies to students. As a class, choose a place to sit and journal. Brainstorm journaling ideas related to bodies of water. For journaling at the kindergarten to grade-two level, students may wish to:

- sketch or colour the water they see around them
- identify emotions, recording feelings about the water (students may use happy faces, emojis, or their own designs for emotions)
- record sounds coming from the water, using pictures and words
- record movements in the water, using pictures and words

Portage & Main Press, 2019 · *Hands-On Science for British Columbia · Land, Water, and Sky for Grades K–2* · ISBN: 978-1-55379-797-5

Portage & Main Press, 2019 · Hands-On Science for British Columbia · Land, Water, and Sky for Grades K–2 · ISBN: 978-1-55379-797-5

14

- take a photo that shows characteristics of the water and sketch it
- draw labelled diagrams showing features of the water
- use all their senses to describe the water using pictures and/or words
- write a poem about what they observe or feel about water

Photograph students' examples for use back in the classroom. Students can also collect water samples for further investigation in the classroom.

Ask a local Elder or Knowledge Keeper to guide the nature walk. They can share information and stories about water and related topics such as rivers, lakes, oceans, fish, and creation stories.

NOTE: See Indigenous Perspectives and Knowledge, page 9, for guidelines for inviting Elders and Knowledge Keepers to speak to students.

Explore Part Two

Back in the classroom, have students share their photographs and journal entries.

Also share Indigenous stories about water. For example:

- *Creation Story: Sky Woman* by Michelle Corneau. This Kanyen'kehà:ka story is about how the world was once just water, and tells of the origin of the land, when the world was once just water. The Kanyen'kehà:ka is one of Six Nations that together are the Haudenosaunee.
- *From the Mountains to the Sea: We Are a Community* by Brenda Boreham and Terri Mack tells the story of the origins and life of a river.

Explore Part Three

Have students brainstorm types of water found on Earth. Record their ideas on chart paper under the title "Water, Water, Everywhere!" Encourage students to think of as many bodies of water as possible (e.g., ocean, sea, lake, glacier, pond, marsh, river, stream, waterfall, puddle).

Display the map of British Columbia, along with the printed images from the Image Bank: Forms of Water on Earth. Use the map and pictures to encourage further brainstorming and add to the list.

Now, extend the activity by asking:

- Where do you think the water in lakes, rivers, and puddles come from?
- Is rain the only kind of water that falls into the lakes and rivers?

Students will likely suggest other types of precipitation (e.g., snow, ice, sleet, hail, frost, dew, clouds, fog). Add these words to the chart.

Explore Part Four

Examine the map of British Columbia again. First, have students identify their own community, as well as other familiar communities (e.g., places that they lived in the past, communities where other family members live). Have students locate and name a variety of bodies of water in and around their own community, as well as in different regions of the province. Have them share background knowledge and experiences related to these bodies of water. Ask:

- Where are the cities and towns located in relation to these bodies of water?
- Why do you think cities and towns are located near bodies of water?
- Why would people have lived near water long ago?

14

- How are these bodies of water important to Indigenous peoples in British Columbia?

Encourage discussion as to how water is used by people today and how it was used long ago for daily life and survival.

Expand

Provide students with an opportunity to explore forms of water on Earth further by posing their own questions for individualized inquiry. They may wish to:

- Initiate a project at the Makerspace, such as designing and constructing a boat that floats, or a model of a local lake or river.
- Create a graphic organizer to display diagrams and photographs of various forms of water on Earth.
- Research a specific body of water (see page 26 for more information about inquiry through research).
- Explore the properties of water at the water table.
- Conduct an investigation or experiment based on personal inquiry.

As students explore and select ideas to expand learning, provide support and guidance as needed, and offer access to materials and resources that will enable students to conduct their chosen investigations.

Embed Part One: Sharing Circle

Revisit the guided inquiry question: **What forms of water are found on Earth?** Have students share their knowledge, provide examples, and ask further inquiry questions.

Embed Part Two

- Add to concept web as students learn new concepts, answer some of their own inquiry questions, and ask new inquiry questions.

- Add new terms and illustrations to the class word wall. Include the words in languages other than English, as appropriate.
- Focus on students' use of the Core Competencies. Have students reflect on how they used one of the Core Competencies (Thinking, Communicating, or Personal and Social Skills) during the various lesson activities. Project one of the CORE COMPETENCY DISCUSSION PROMPTS templates (pages 38–42), and use it to inspire group reflection. Referring to the template, choose one or two "I Can" statements on which to focus. Students then use the "I Can" statements to provide evidence of how they demonstrated that competency. Ask questions directly related to that competency to inspire discussion. For example:
  - How do you feel about the bodies of water we explored on our nature walks? (Personal Awareness and Responsibility)

  Have students reflect orally, encouraging participation, questions, and the sharing of evidence. (See page 29 for more information on these templates.)

  As part of this process, students can also set goals. For example, ask:
  - What would you do differently next time and why?
  - How will you know if you are successful in meeting your goal?
- To encourage self-reflection, provide prompts that students can use to cite examples of how they have used the Core Competencies in their learning. For this purpose, the CORE COMPETENCY SELF-REFLECTION FRAMES (pages 43–47) can be used throughout the learning process. There are five frames provided to address the Core Competencies: Communication, Creative Thinking, Critical Thinking, Positive Personal and Cultural Identity, and Personal Awareness and Responsibility. Teachers can conference

▶

Portage & Main Press, 2019 · *Hands-On Science for British Columbia · Land, Water, and Sky for Grades K–2* · ISBN: 978-1-55379-797-5

14

individually with students to support self-reflection, or students may complete prompts using words and pictures.

Again, have students set goals by considering what they might do differently on future tasks and how they will know if they are successful in meeting their goal.

NOTE: Use the same prompts from these templates over time to see how thinking changes with different activities.

Enhance

- **Family Connections**: Provide students with the following sentence starter:
  - My family's favourite body of water is _____.

  Have students complete the sentence starter at home. Family members can help students draw and write about this topic. Have students share their sentences with the class.

- Learn about the Whanganui River in New Zealand, which was given the same rights as humans, as it has been recognized as an ancestor. This is reflective of the perspective of Indigenous peoples in New Zealand that inanimate objects in nature have life and all nature is connected. Go to <https://www.theguardian.com/world/2017/mar/16/new-zealand-river-granted-same-legal-rights-as-human-being>.

Portage & Main Press, 2019 · *Hands-On Science for British Columbia · Land, Water, and Sky for Grades K–2* · ISBN: 978-1-55379-797-5

15 | How Does Water Move Through the Water Cycle?

Information for Teachers

Not only does water come in many forms, it can easily change from one form to another. This is called *changing states*. Water changes from liquid (water in a pond) to solid (ice on surface of pond) by the change of state called *freezing*; water (in a kettle) changes from liquid to gas (water vapour) due to *evaporation*; and water changes from a solid (ice cube) to liquid due to *melting*. On Earth, water changes state frequently—ice, liquid water, water vapour— and the results are seen as dew, rain, water in lakes/rivers/streams/creeks/ponds, frost, snow, sleet, hail, or fog/ice crystals (in the air). These changes are part of a *water cycle* whereby water is never really lost but, rather, simply changes from one form to another.

An exploration of the water cycle is another opportunity to focus on sustainability, in that water is not lost or gained but changes state.

Materials

- chart paper
- markers
- digital camera
- electric kettle or pot and hotplate
- jug of water
- food colouring
- zipper-lock bags (one for each student)
- warm water
- graduated beaker or measuring cup
- aluminum pie plate
- ice cubes
- Image Bank: Mountains (see Appendix, page 165)
- Diagram: The Water Cycle (4.15.1)
- concept web (from lesson 3)

Engage

Lead students in recreating the sounds of a rainstorm. Have them seated in chairs for this activity. Demonstrate each action, and have students replicate the action to create the progressive sounds of rain. Say:

- The rainstorm begins with drizzle. (*Rub hands together, palms and fingers flat, back and forth slowly, then faster.*)
- Big raindrops begin to fall. (*Snap your fingers slowly, then faster.*)
- The rain begins to pour down heavily. (*Quickly pat your lap with your hands, faster, then faster.*)
- The rain is really pouring down now. (*Stamp feet on the floor.*)
- The storm begins to lessen. (*Stop stamping your feet and go back to patting your lap with your hands.*)
- It lessens more. (*Stop lap patting, and snap your fingers again.*)
- The rain begins to let up. (*Slow the finger snapping, then begin "drizzle" again by rubbing your hands together.*)
- Finally, the rain stops. (*Stop all actions.*)

As a class, brainstorm a list of songs, poems, or rhymes about rain and snow. For example:

- "It's Raining, It's Pouring"
- "Itsy Bitsy Spider"
- "Rain, Rain, Go Away"
- "Raindrops Keep Falling on My Head"
- "Let It Snow"
- "Let It Go" (from Frozen) ("The snow glows white on the mountain tonight")
- "Jingle Bells" (Dashing Through the Snow)

Sing the songs, and have students identify the words that have to do with forms of water. Record these words on chart paper.

Introduce the guided inquiry question: **How does water move through the water cycle?**

Portage & Main Press, 2019 · *Hands-On Science for British Columbia · Land, Water, and Sky for Grades K–2* · ISBN: 978-1-55379-797-5

Portage & Main Press, 2019 · Hands-On Science for British Columbia · Land, Water, and Sky for Grades K–2 · ISBN: 978-1-55379-797-5

15

Explore Part One

 SAFETY NOTE: For safety reasons, this activity is best done as a demonstration. Let students participate in certain aspects of the demonstration, as noted below.

NOTE: If possible, have students photograph each stage of the demonstration.

Have students gather around the demonstration table. Discuss precipitation. Ask:

- What happens to make rain fall from the sky?

Discuss students' conceptions of precipitation, noting their understanding of the concept. Explain that you are going to build a model to demonstrate how rain is made. Explain to students that they are going to observe what needs to happen for precipitation to occur.

Have a volunteer student use a graduated beaker or measuring cup to measure 250 mL of water and pour it into the kettle or pot. Explain that this water represents water in lakes, rivers, and other bodies of water, as well as water in the ground and on plants and other surfaces, such as dew. Ask:

- What do you think happens to water in these places when the Sun shines on it?

Discuss that warmth from the Sun heats the water.

Now, use the kettle (or pot and hot plate) to heat the 250 mL water, to simulate the sun heating water on Earth. Have students observe the water as it is heated. Ask:

- What do you observe?
- What is happening to the heated water?
- Write the term *evaporation* on chart paper. With students, co-construct a definition and have students check their dictionaries. Then, discuss *evaporation* focusing on the change of water from a liquid to a gas when it is

heated. Since water vapour is lighter than liquid water, it rises. Explain that this is what happens to water in rivers and lakes when it is heated by the Sun. Ask:

- What proof do we have that water evaporates from the small streams and creeks in our fields? (They can dry up as the season progresses.)

Explain how this water evaporates and rises into the sky. Ask:

- What do you think the temperature is like high up in the sky?
- Is it very hot or very cold?

Display the Image Bank: Mountains. Examine and discuss photographs of snow-covered mountains, with greenery at the base and snow at the top. Ask:

- What do these images tell you about temperatures high up and on the ground?
- Why do you think temperatures are colder higher up and warmer near the Earth's surface?

Discuss students' ideas. Explain that the temperature above land is cold, since heat from the sun is absorbed by and reflected off the land, warming the area closest to the ground. Simulate the cold area above the land by holding an aluminum pie plate filled with ice cubes above the evaporating water from the kettle or pot.

Have students observe the bottom of the pie plate. Ask:

- What do you notice on the bottom of the pie plate?
- Where are these water droplets coming from?
- What is happening to the droplets as they get bigger?

Explain that as the water vapour is cooled by the ice cubes in the plate, it condenses into liquid form as water droplets. When these droplets run into one another, they join together and get bigger and heavier, eventually falling from the pie plate. This is what happens to water vapour in the sky. The water vapour cools, condenses, and forms droplets that fall to the ground as rain. This complete process is called the *water cycle*.

Write the term *condensation* on chart paper. With students, co-construct a definition and have students check their dictionaries. In their own words, have students describe the water cycle using the terms *condensation* and *evaporation* (or *condense* and *evaporate*).

Refer back to the discussion about streams and creeks drying up. Ask:

■ How do dry stream and creek beds get refilled with water? (rain, run off from surrounding land)

Discuss how, in winter, the outside air temperature is so cold the water droplets in the air freeze into snow. Ask:

■ In the winter, do the droplets fall as rain?

■ What happens to that snow as air temperatures rise in the spring?

■ What are the effects of melting snow—what happens when snow melts? (puddles, high water in rivers and streams, wet soil, sometimes flooding)

Explore Part Two

Explain to students that they are going to make individual water-cycle models.

NOTE: Young students may require adult or older peer support.

Distribute one zipper-lock bag to each student. Have students pour 125 mL of warm water, with a drop of food colouring, into the bag, zip it closed, and tape it to the window. Tell them to watch for "clouds and rain" as in the following examples:

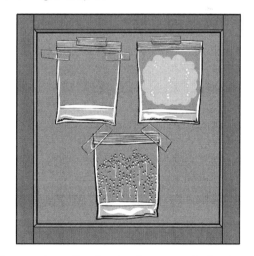

As condensation begins to appear in the bags, discuss the water cycle as a class. Then, have students use permanent markers to label their water cycle models. After students have constructed and labelled their models, display the Diagram: The Water Cycle (4.15.1):

Portage & Main Press, 2019 · *Hands-On Science for British Columbia · Land, Water, and Sky for Grades K–2* · ISBN: 978-1-55379-797-5

▶

Portage & Main Press, 2019 · *Hands-On Science for British Columbia · Land, Water, and Sky for Grades K–2* · ISBN: 978-1-55379-797-5

15

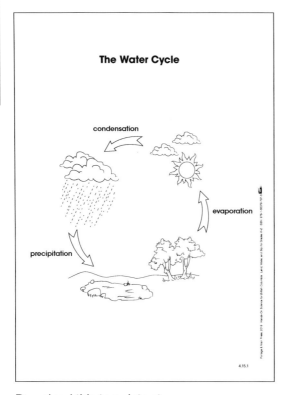

The Water Cycle

condensation

evaporation

precipitation

4.15.1

Download this template at
<www.portageandmainpress.com/product/
HOSLandWaterSkyK2>.

Have students describe the process in their own words, using their models for reference. Also have them compare and match the parts of their diagrams to the Diagram: The Water Cycle.

Expand

Provide students with an opportunity to explore the water cycle further by posing their own questions for individualized inquiry. They may wish to:

- Initiate a project at the Makerspace, such as designing and constructing a unique model of the water cycle.
- Research to find out how hail and sleet are formed, and create a presentation for the class (see page 26 for more information about inquiry through research).

- Research various types of snowflakes and find out why no two snowflakes are alike.
- Write poems and songs about the water cycle.
- Conduct an investigation or experiment based on their own inquiry questions.

As students explore and select ideas to expand learning, provide support and guidance as needed, and offer access to materials and resources that will enable students to conduct their chosen investigations.

Student Self-Assessment Ⓒ

Distribute the SCIENCE JOURNAL template, on page 37, to each student, and have students reflect on their learning about the water cycle. The focus here is on sharing observations through written language and drawing.

Embed Part One: Sharing Circle

Revisit the guided inquiry question: **How does water move through the water cycle?** Have students share their experience and knowledge, provide examples, and ask further inquiry questions.

Embed Part Two

- Add to the concept web as students learn new concepts, answer some of their own inquiry questions, and ask new inquiry questions.
- Add new terms and illustrations to the class word wall. Include the words in languages other than English, as appropriate.
- Focus on students' use of the Core Competencies. Have students reflect on how they used one of the Core Competencies (Thinking, Communicating, or Personal and Social Skills) during the various lesson activities. Project one of the CORE COMPETENCY DISCUSSION PROMPTS templates (pages 38–42), and use it to inspire group

Hands-On Science for British Columbia

reflection. Referring to the template, choose one or two "I Can" statements on which to focus. Students then use the "I Can" statements to provide evidence of how they demonstrated that competency. Ask questions directly related to that competency to inspire discussion. For example:

- How did you share your learning today? (Communication)

Have students reflect orally, encouraging participation, questions, and the sharing of evidence. (See page 29 for more information on these templates.)

As part of this process, students can also set goals. For example, ask:

- What would you do differently next time and why?
- How will you know if you are successful in meeting your goal?

- To encourage self-reflection, provide prompts that students can use to cite examples of how they have used the Core Competencies in their learning. For this purpose, the CORE COMPETENCY SELF-REFLECTION FRAMES (pages 43–47) can be used throughout the learning process. There are five frames provided to address the Core Competencies: Communication, Creative Thinking, Critical Thinking, Positive Personal and Cultural Identity, and Personal Awareness and Responsibility. Teachers can conference individually with students to support self-reflection, or students may complete prompts using words and pictures.

Again, have students set goals by considering what they might do differently on future tasks and how they will know if they are successful in meeting their goal.

NOTE: Use the same prompts from these templates over time to see how thinking changes with different activities.

Enhance

- **Family Connections**: Provide students with the following sentence starter:
  - A family story about rain or snow is

    _____.

  Have students complete the sentence starter at home. Family members can help students draw and write about this topic. Have students share their sentences with the class.
- Have students work in groups to act out the water cycle.
- Explore a different model of the water cycle. Distribute one clear plastic cup to each student. Have students put some warm coloured water into the cup, and then cover it with plastic wrap. Next, have them put ice cubes on top of the plastic wrap, and place the model near a heat source, such as a heat vent or radiator. Tell them to watch for "clouds and rain."

- Use a rain stick to explore the sounds that rain makes. Have students participate in a talking circle by passing around the rain stick. When it is their turn to talk, they may share an experience they had with rain or a rainstorm.
- Leave a bowl or plate of water on a window sill and allow it to evaporate, as students

Portage & Main Press, 2019 · *Hands-On Science for British Columbia · Land, Water, and Sky for Grades K–2* · ISBN: 978-1-55379-797-5

Portage & Main Press, 2019 · Hands-On Science for British Columbia · Land, Water, and Sky for Grades K–2 · ISBN: 978-1-55379-797-5

15

observe changes. On a window sill, a shallow layer of water could be photographed over time as it disappears through evaporation. Use a device with a timer such as an iPad or other device with iMotion to create a time lapse.

- Explore demonstrations of the water cycle online. Go to websites such as:
  - "Watershed Basics: The Water Cycle" go to <https://www.crd.bc.ca/education/our-environment/watersheds/watershed-basics/water-cycle>
  - "NASA Science—Water Cycle" go to <science.nasa.gov/earth-science/oceanography/ocean-earth-system/ocean-water-cycle>
- Have students explore the water cycle over the course of a year, to see the variety of forms of water. Visit Snowmelt—The Water Cycle. Go to: <water.usgs.gov/edu/watercyclesnowmelt.html>.
- Conduct further investigations with students that focus on the water cycle. For example:
  - **Condensation:** Moisture in the air condenses to a liquid form as it cools or comes in contact with a cool surface. Some examples of this are:
    - touch a mirror to the tip of your nose and exhale
    - breathe onto the lenses of a pair of eye glasses
    - place ice cubes in a glass of water and let the glass and ice cubes stand at room temperature (small water droplets will form on the outside of the glass as the air around it cools)
  - **Water vapour:** Water vapour is water in its gaseous state—small particles of water are suspended in the air. Some examples are:
    - your breath on a cold day
    - steam from boiling water

- **Clouds:** Explain that cloud formation is similar to seeing your breath on a cold day: warm, moist air from your lungs cools as it meets the outside cold air, causing the air to condense and become visible, just like the clouds in the sky.
- **Precipitation**: Discuss how other forms of precipitation (e.g., snow, hail, sleet) are formed.
- **Snow**: Snow is not frozen rain. Snow begins as tiny ice crystals that form when water vapour chills and crystallizes. As more water vapour rises into the air, more water is deposited on the ice crystals and the crystals grow. Soon, the larger crystals fall as snow.
- **Hail**: Hailstones are formed when water in the top of cumulonimbus clouds freezes. As the hail falls, it collects more water and grows, building up layer on layer of ice.
- **Sleet**: Sleet occurs when a layer of warm air lies above a layer of cold air. Snow falling from higher clouds melts as it passes through the warm layer, and it turns to rain. This rain then continues to fall through the cold layer of air, where it freezes and turns to pellets of ice.

16 | What Are Sources of Safe Drinking Water?

Information for Teachers

Water distribution systems take water from the source to people's homes. In Canada, distribution systems include pumps, pipes, storage tanks, reservoirs, and underground storage tanks. In many cases, the water passes through a water treatment and testing process to ensure the water entering our homes is clean.

NOTE: Many water distribution plants do not provide tours. As a result, this lesson relies on guest speakers and virtual tours.

About one-fifth of all First Nations communities in Canada face boil water advisories and many have no access to running water at all. For more information, see the following resources:

- <https://www.canada.ca/en/indigenous-services-canada/services/short-term-drinking-water-advisories-first-nations-south-60.html>
- <www.fnha.ca/what-we-do/environmental-health/drinking-water-advisories>
- <https://www.cbc.ca/news/canada/british-columbia/first-nations-water-solutions-1.3482568>
- <https://www.theglobeandmail.com/news/british-columbia/more-drinking-water-advisories-for-bc-than-any-other-province-report-finds/article23488553/>

Materials

- *The Magic School Bus at the Waterworks* by Joanna Cole
- pictures of the local water distribution system (posters and brochures are often available from your local water company)
- chart paper
- markers
- provincial map
- computer/tablet with internet access
- Learning-Centre Task Card: How Can We Construct a Water Filtration Device? (4.16.1)
- writing or drawing paper
- 2-L pop bottles (with the bottoms cut off)
- muddy or oily water
- charcoal
- fine sand
- coarse sand
- pebbles
- cotton wool
- bowl
- beaker or measuring cup
- digital camera
- concept web (from lesson 3)

Engage

As a class, brainstorm ways humans use water. On chart paper, make a list of students' ideas. Include uses of water in the classroom, in the rest of the school, at home, in the neighbourhood, city, province, country, and the world. Ask:

- Where do you think your drinking water comes from? (e.g., tap, well, lake.)

NOTE: Be sure to keep the chart paper list of students' ideas of how humans use water, as it is used again in the lesson 17.

Introduce the guided inquiry questions: **What are sources of safe drinking water?**

Explore Part One

Read *The Magic School Bus at the Waterworks* by Joanna Cole. Discuss the story, and relate it to your local area.

Use the provincial map to locate the source of water for your community. Trace the route the water takes from the source to your community.

Portage & Main Press, 2019 · *Hands-On Science for British Columbia · Land, Water, and Sky for Grades K–2* · ISBN: 978-1-55379-797-5

Portage & Main Press, 2019 · *Hands-On Science for British Columbia · Land, Water, and Sky for Grades K–2* · ISBN: 978-1-55379-797-5

16

NOTE: In some cases water does not pass through a treatment centre but still needs to be monitored. For example, the water in rural wells, such as on farms and reserves, is not treated. Property owners must have their water tested regularly to ensure it is safe for drinking and for crops.

Explore Part Two

As a class, explore the issue of communities in British Columbia without access to safe drinking water. Access the links under **Information for Teachers** to identify these communities. Internet searches will also highlight media coverage of some of these communities and the challenges they face. As a class, watch news clips to better understand local issues.

NOTE: This may be a sensitive issue for students, families, and communities dealing with concerns around safe drinking water. Approach this topic with care and compassion.

Explore Part Three

Invite a guest speaker from a local water purification plant or water company, or an expert on local water sources (and/or issues) to present to the class. Or explore virtual tours of the plant with students online. Before the presentation/virtual tour, have students brainstorm a list of questions they have about how water is distributed in their community. Record their questions on chart paper, so students have an opportunity to have their questions answered.

NOTE: Use this opportunity to ask questions or learn more about communities lacking safe drinking water.

After the presentation/virtual tour, have a class discussion about the reasons why water needs to be treated. Ask:

- Why is it important to treat water before using it?

- What might be in the water that needs to be removed?
- Is water safe to use just because it looks clear?
- Could the water still be unsafe to drink?

Discuss invisible toxins that can be found in some water sources and how these are removed through treatment processes.

Expand

Provide students with an opportunity to explore sources of safe drinking water further by posing their own questions for individualized inquiry. They may wish to:

- Initiate a project at the Makerspace, such as designing and constructing a model of a unique water filtration systems.
- Research to find out how water is purified in nearby communities (see page 26 for more information about inquiry through research).
- Write poems and songs about safe drinking water.
- Create a how-to picture book about cleaning water.
- Explore various ways to clean/filter water at the water table.
- Conduct an investigation or experiment based on their own inquiry questions.

As students explore and select ideas to expand learning, provide support and guidance as needed, and offer access to materials and resources that will enable students to conduct their chosen investigations.

Learning Centre

At the learning centre, provide 2-L pop bottles with the bottoms cut off, muddy or oily water, charcoal, fine sand, coarse sand, pebbles, cotton wool, a bowl, and a beaker or measuring cup. Also, provide paper for recording observations and a copy of the Learning-Centre Task Card: How Can We Construct a Water Filtration Device? (4.16.1)

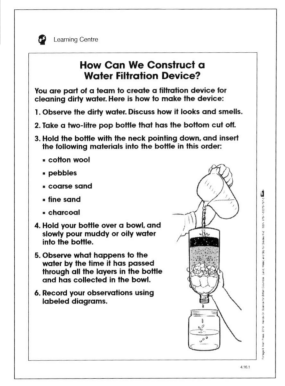

Learning Centre

How Can We Construct a Water Filtration Device?

You are part of a team to create a filtration device for cleaning dirty water. Here is how to make the device:

1. Observe the dirty water. Discuss how it looks and smells.

2. Take a two-litre pop bottle that has the bottom cut off.

3. Hold the bottle with the neck pointing down, and insert the following materials into the bottle in this order:

- cotton wool
- pebbles
- coarse sand
- fine sand
- charcoal

4. Hold your bottle over a bowl, and slowly pour muddy or oily water into the bottle.

5. Observe what happens to the water by the time it has passed through all the layers in the bottle and has collected in the bowl.

6. Record your observations using labeled diagrams.

4.16.1

Download this template at <www.portageandmainpress.com/product/HOSLandWaterSkyK2>.

In small working groups, have students construct water filtration devices for cleaning dirty water.

 SAFETY NOTE: Make sure students understand that although the water is now cleaner, it may not be safe to drink.

Formative Assessment (E)

Observe students as they work in groups to construct their water filtration systems. Focus on their ability to compare observations with those of other students. Use the COOPERATIVE SKILLS TEACHER ASSESSMENT template, on page 52, to record results. Provide descriptive feedback to

students about how they collaborate with others. Be sure to document student progress with videos or photographs as evidence of learning.

Student Self-Assessment (AI)

Have students complete the COOPERATIVE SKILLS SELF-ASSESSMENT template, on page 49, to reflect on their work. This reflection focuses on students taking part in caring for self, family, classroom and school.

Embed Part One: Sharing Circle

Revisit the guided inquiry question: **What are sources of safe drinking water?** Have students share their experiences and knowledge, provide examples, and ask further inquiry questions.

Embed Part Two

- Add to the concept web as students learn new concepts, answer some of their own inquiry questions, and ask new inquiry questions.

- Add new terms and illustrations to the class word wall. Include the words in languages other than English, as appropriate.

- Focus on students' use of the Core Competencies. Have students reflect on how they used one of the Core Competencies (Thinking, Communicating, or Personal and Social Skills) during the various lesson activities. Project one of the CORE COMPETENCY DISCUSSION PROMPTS templates (pages 38–42), and use it to inspire group reflection. Referring to the template, choose one or two "I Can" statements on which to focus. Students then use the "I Can" statements to provide evidence of how they demonstrated that competency. Ask questions directly related to that competency to inspire discussion. For example:

  - How did you decide which questions to ask today? (Critical Thinking)

▶

Portage & Main Press, 2019 · *Hands-On Science for British Columbia · Land, Water, and Sky for Grades K–2* · ISBN: 978-1-55379-797-5

Have students reflect orally, encouraging participation, questions, and the sharing of evidence. (See page 29 for more information on these templates.)

As part of this process, students can also set goals. For example, ask:

- What would you do differently next time and why?
- How will you know if you are successful in meeting your goal?

■ To encourage self-reflection, provide prompts that students can use to cite examples of how they have used the Core Competencies in their learning. For this purpose, the CORE COMPETENCY SELF-REFLECTION FRAMES (pages 43–47) can be used throughout the learning process. There are five frames provided to address the Core Competencies: Communication, Creative Thinking, Critical Thinking, Positive Personal and Cultural Identity, and Personal Awareness and Responsibility. Teachers can conference individually with students to support self-reflection, or students may complete prompts using words and pictures. Again, have students set goals by considering what they might do differently on future tasks and how they will know if they are successful in meeting their goal.

NOTE: Use the same prompts from these templates over time to see how thinking changes with different activities.

Enhance

- **Family Connections:** Provide students with on of the following sentence starters:
  - My family gets our drinking water from _____.
  - When we go camping, we get our drinking water from _____.

Have students complete the sentence starter at home. Family members can help students draw and write about this topic. Have students share their sentences with the class.

■ Discuss students' camping experiences, and have them share how they accessed clean water. Also discuss how people who go backcountry camping (in the undeveloped wilderness, where there are no campgrounds and which cannot be easily accessed by vehicle) find and clean water for drinking.

■ Discuss with students that if a water filtration system malfunctions, water must be boiled before it is safe to drink. Students may have heard about (or had experience with) a "boil water advisory." Visit the Health Canada page "Boil Water Advisories and Boil Water Orders" for more information. Go to: <www.hc-sc.gc.ca/ewh-semt/pubs/water-eau/boil-ebullition-eng.php>.

Focus on positive changes and solutions—communities that did not have access to clean water and now do, and how some communities solved this problem.

■ Discuss the use of various water filtration and/or cleaning systems found right in homes and businesses. The intent is not to create a lengthy list or to understand the differences between these, but rather to recognize there are many ways to treat water.

Portage & Main Press, 2019 · Hands-On Science for British Columbia · Land, Water, and Sky for Grades K–2 · ISBN: 978-1-55379-797-5

17 | How Can We Use Water Wisely?

Information for Teachers

Each Canadian uses approximately 300–350 litres of water *every day* for drinking, cleaning, cooking, watering gardens, and flushing away waste.

Although water resources on Earth appear to be plentiful, suitable water available for daily use by humans is limited. Water resources are finite and must be conserved in order to ensure there will still be plenty of water for all humans in years to come.

Although water is essential to life, in many parts of the world even tap water can be unsafe to drink. In some places, reliable and safe water sources are extremely scarce. For example, in areas affected by drought, people must travel long distances with buckets or vessels to collect water at specific times, and then store the water safely.

In other places where clean water is limited, scientists and engineers have developed technology to provide safe and reliable supplies of water. But, some of these technologies, such as desalination plants, are extremely expensive.

Many Indigenous peoples recognize that water is sacred as the giver of life, interconnected to all life—human, animal, and plant—and it is important to protect our water from pollution and overconsumption. Some reserves in British Columbia are under either permanent boil-water advisories, or have contaminated water. A list of these locations can be found here: <http://www.fnha.ca/what-we-do/environmental-health/drinking-water-advisories>

Materials

- chart-paper list of students' ideas of how humans use water (from lesson 16)
- Chart: Average Water Usage (4.17.1)
- variety of 1-L containers of different shapes and made from various materials (e.g., pop bottles, milk cartons, juice containers)
- 4-L container
- pail of water
- calculators (optional)
- chart paper
- markers
- Template: Daily Water Usage in My Home (4.17.2)
- Learning-Centre Task Card: How Can We Use Water Wisely? (4.17.3)
- *The Water Walker* by Joanne Robertson
- writing or drawing paper
- concept web (from lesson 3)

Engage

Display the chart-paper list of students' ideas of how humans use water (from lesson 16). Have students add to the list any new ideas they have about how humans use water. Ask:

- How much water do you think you use in your home every day?

Have students examine a variety of 1-L containers. Ask:

- How many litres of water do you think you use in a day?
- How much do you think you use for drinking?
- What about for showering or bathing? Brushing your teeth? Cleaning and laundry? Washing fruit, vegetables, and dishes? Flushing the toilet?

Have students predict how many litres of water they use in a day. List all students' predictions on chart paper. Then, help each student multiply their number prediction by the number of people in their family to come up with a daily family water-use prediction. Record this next to each student's personal prediction.

Portage & Main Press, 2019 · *Hands-On Science for British Columbia · Land, Water, and Sky for Grades K–2* · ISBN: 978-1-55379-797-5

Introduce the guided inquiry question: **How can we use water wisely?**

Explore Part One

Display the Chart: Average Water Usage (4.17.1):

Average Water Usage

Activity	Usage
brushing teeth, water on	6 litres
brushing teeth, water off	1 litre
bath	80 litres
shower (per minute)	8 litres
flushing the toilet	6 litres
washing dishes by hand	25 litres
dishwasher	21 litres
one load of laundry	150 litres
drinking and cooking	15 litres (per person per day)
watering the lawn	450 litres
washing the car	35 litres

4.17.1

Focus on how much water is used to brush teeth.

Have a student fill the four-litre container with water to show how much water is used to brush teeth when the tap is left on while brushing. Have another student fill one of the one-litre containers with water to show how much water is used to brush teeth when the tap is turned off while brushing.

Download this template at <www.portageandmainpress.com/product/HOSLandWaterSkyK2>.

NOTE: After the demonstration, use the water from the jugs to water plants in the classroom or school.

Explain that turning off the tap is one way of *conserving* or reducing the amount of water we use.

Review other data on the chart. Focus on the amount of water used for each activity (e.g., doing a load of laundry, flushing the toilet, having a bath). Encourage students to examine the data and generate questions for classmates to answer. For example:

- Is it better to wash dishes by hand or in the dishwasher?
- Which uses more water, flushing the toilet or brushing your teeth?
- Do you usually have a shower or a bath?
- Which do you think uses more water? Why?

Have students share their ideas.

Explore Part Two

As a family connection, challenge students to determine the amount of water used in their home in a day. Provide each student with a letter to parents/guardians that explains the activity and a copy of the Chart: Average Water Usage (4.17.1).

Discuss with students how to record their data. As a class, you may design your own recording sheet or use the Template: Daily Water Usage in My Home (4.17.2):

Portage & Main Press, 2019 · Hands-On Science for British Columbia · Land, Water, and Sky for Grades K–2 · ISBN: 978-1-55379-797-5

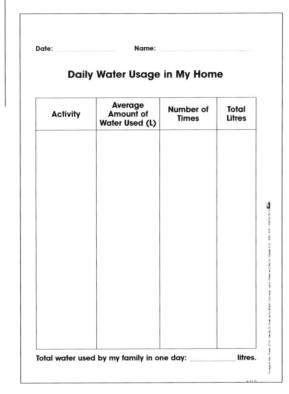

Download this template at
<www.portageandmainpress.com/product/
HOSLandWaterSkyK2>.

Review with students how to record the amount
of water used in their home in one day.

NOTE: Have students collect this data on a
weekend day when they can be at home to
record results. Family members can also help
with this task.

When students have completed the activity
sheet, have them present their findings to the
class. Encourage students to examine the data
and generate questions for each other to answer.

Record the actual amount of water used in each
student's home next to that student's prediction
(from the Engage activity) of how much water
their family uses in a day.

As a class, use this data to create a pictograph
on chart paper. Co-construct criteria for the
pictograph with students. For example:

- create an appropriate title
- label each student
- use pictures for each litre
- count and record correctly

Student Self-Assessment

After students have concluded their home
surveys, have them complete the STUDENT SELF-
ASSESSMENT template, on page 35, to reflect
on their learning about water usage. This task
focuses on considering some environmental
consequences of their actions.

Expand

Provide students with an opportunity to explore
wise water usage further by posing their own
questions for individualized inquiry. They may
wish to:

- Initiate a project at the Makerspace, such as
  examining shower savers and then designing
  and constructing a model shower saver to
  reduce water usage.
- Create a poster to encourage wise use
  of water.
- Make a presentation for parents and
  community members about actions to
  reduce water usage, or present at a school
  assembly.
- Create a picture book about water usage.
- Design and make bookmarks with water
  saving tips on them to share with the public
  or school peers.
- Conduct an investigation or experiment
  based on their own inquiry questions.

As students explore and select ideas to expand
learning, provide support and guidance as
needed, and offer access to materials and
resources that will enable students to conduct
their chosen investigations.

Portage & Main Press, 2019 · Hands-On Science for British Columbia · Land, Water, and Sky for Grades K–2 · ISBN: 978-1-55379-797-5

Learning Centre

At the learning centre, provide a copy of *The Water Walker* by Joanne Robertson and drawing paper, along with a copy of the Learning-Centre Task Card: How Can We Use Water Wisely? (4.17.3):

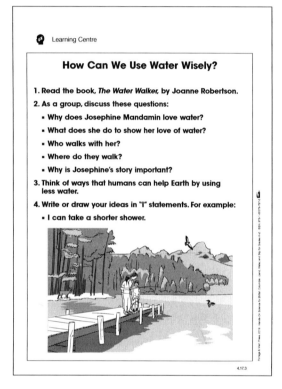

Download this template at <www.portageandmainpress.com/product/HOSLandWaterSkyK2>.

Have students read the book, then respond to questions about the story. Have them think of ways humans can reduce the amount of water they consume each day to help the Earth. On the drawing paper, have students write or draw their solutions in "I" statements. For example:

- I can take showers instead of baths to conserve water.

- I can turn off the water while brushing my teeth.

- I can run washing machines and dishwashers only when they are fully loaded.

- I will check for leaky taps.

- I will check with my parents to find out if the dishwasher has an energy-saving cycle.

- I will add mulch to the garden soils. It helps to keep moisture in the soil.

- I will check to see if our toilet has a water-saving device.

- I will use a broom instead of a hose to clean our driveway.

- I will find the water metre (with my family's help) and learn how to read it.

Embed Part One: Sharing Circle

Revisit the guided inquiry question: **How can we use water wisely?** Have students share their knowledge and experience, provide examples, and ask further inquiry questions.

Embed Part Two

- Add to the concept web as students learn new concepts, answer some of their own inquiry questions, and ask new inquiry questions.

- Add new terms and illustrations to the class word wall. Include the words in languages other than English, as appropriate.

- Focus on students' use of the Core Competencies. Have students reflect on how they used one of the Core Competencies (Thinking, Communicating, or Personal and Social Skills) during the various lesson activities. Project one of the CORE COMPETENCY DISCUSSION PROMPTS templates (pages 38–42), and use it to inspire group reflection. Referring to the template, choose one or two "I Can" statements on which to focus. Students then use the "I Can" statements to provide evidence of how they demonstrated that competency. Ask questions directly related to that competency to inspire discussion. For example:

▶

Portage & Main Press, 2019 · *Hands-On Science for British Columbia · Land, Water, and Sky for Grades K–2* · ISBN: 978-1-55379-797-5

17

- How did you grow as a learner today? (Positive Personal and Cultural Identity)

Have students reflect orally, encouraging participation, questions, and the sharing of evidence. (See page 29 for more information on these templates.)

As part of this process, students can also set goals. For example, ask:

- What would you do differently next time and why?
- How will you know if you are successful in meeting your goal?

- To encourage self-reflection, provide prompts that students can use to cite examples of how they have used the Core Competencies in their learning. For this purpose, the CORE COMPETENCY SELF-REFLECTION FRAMES (pages 43–47) can be used throughout the learning process. There are five frames provided to address the Core Competencies: Communication, Creative Thinking, Critical Thinking, Positive Personal and Cultural Identity, and Personal Awareness and Responsibility. Teachers can conference individually with students to support self-reflection, or students may complete prompts using words and pictures. Again, have students set goals by considering what they might do differently on future tasks and how they will know if they are successful in meeting their goal.

NOTE: Use the same prompts from these templates over time to see how thinking changes with different activities.

Enhance

- **Family Connections**: Provide students with the following sentence starter:
  - In our home, we try to use water wisely by _____.

Have students complete the sentence starter at home. Family members can help students

draw and write about this topic. Have students share their sentences with the class.

- After students have looked at their current water usage and learned about some ways to use water wisely, have students share their initial graph/survey results with their family, along with ways to be wise with water. Then have them conduct a second home survey, to determine if there has been any change.

- Create bar graphs to show the amount of water used for various activities within one household, or select one specific activity and make a graph comparing water used in various households.

- Every summer, many communities in British Columbia are under water restrictions (e.g., do not wash cars and decks, water plants only at certain times of day). Explore this issue with students to see how wise water use is considered at a larger level (e.g., community, city, or municipality).

- Write and illustrate a class big book titled *Water Saving Tips*, in which every student creates a page. The pages can be bound and the book can be donated to the school or local library.

- Compare water usage in households in various countries (e.g., Canadians use about 350 litres of water per day, Britons use an average of 175 litres per day, Bangladeshies use about 45 litres per day).

- Ask students if their families take advantage of rain water by capturing water in a rain barrel or maintaining a rain garden. Discuss how relatively clean rain water can be, especially for use in gardens.

- Read books about how families around the world collect water for daily use. Focus on areas where water is not as accessible as it is in many communities in Canada. Tell students that many people in the world spend as many as six hours each day collecting water for their family. Ask:

Portage & Main Press, 2019 · *Hands-On Science for British Columbia · Land, Water, and Sky for Grades K–2* · ISBN: 978-1-55379-797-5

17

- How would your life change if you had to spend six hours every day collecting water?
- Is there anything you would change about the way you use water now if you had to walk long distances to collect the water?

■ Have students play a bucket-relay game, to help them understand how precious water becomes for those who have to walk long distances to get it. Organize the class into teams; each team has two buckets—one full of water, the other empty. The object is for teams to use plastic cups to carry water from one bucket to another, in an attempt to be first to fill their bucket at the finish line. A variation of the game is to have students use cups with a pinhole in the bottom to make the task more difficult. Note that this relay activity is best conducted outside.

Portage & Main Press, 2019 · *Hands-On Science for British Columbia · Land, Water, and Sky for Grades K–2* · ISBN: 978-1-55379-797-5

18 | How Can We Keep Our Water Clean?

Information for Teachers

Water is essential to all plant and animal life on Earth. Clean water is plentiful in some countries. In many countries, however, clean, safe water is not always available for drinking or for adequate sanitation. Twenty-five million people die each year due to a lack of water, sanitary facilities, and from waterborne diseases; three-fifths of these people are children. In some countries, the main source of drinking water may be contaminated when raw sewage enters ponds and rivers and/or when water sources are used for laundry. These unsanitary conditions can breed deadly waterborne diseases. Half of the world's leading diseases breed in water or are spread by water.

Even in countries where water is plentiful, scientists are finding that much of it is polluted. Groundwater can be contaminated by chemicals, pesticides, and run-off from landfills. Surface waters can also be contaminated by poisonous industrial chemicals and untreated sewage. Even in Canada, which has more fresh water than most countries, many First Nations communities are unable to drink their water because of pollution caused by human activity.

Do not despair! We can reduce pollution and conserve the resources that we have. Children and adults alike must do their part.

Materials

- *Salmon Boy: A Legend of the Sechelt People* by Donna Joe
- *A River Ran Wild: An Environmental History* by Lynne Cherry
- chart paper
- markers
- books about water pollution, especially ones that focus on oil spills
- vegetable oil
- measuring cup
- water
- 2-litre plastic pop bottle with cap
- food colouring (any colour)
- digital camera
- dishpan or basin
- feathers
- small pieces of faux fur
- students' suggested materials for cleaning up an oil spill (see Explore Part Two)
- shoeboxes
- variety of art supplies for building dioramas (e.g., cardboard boxes and tubes, craft sticks, sheets of cardboard or cardstock, wood and fabric scraps, construction paper, scrap paper, glue, tape, scissors)
- Plasticine
- concept web (from lesson 3)

Engage

Read *Salmon Boy: A Legend of the Sechelt People* by Donna Joe. After reading, ask:

- What lessons did you learn from the story?
- Why is the natural world important?
- Why should we care for the environment?

Have students share their ideas. Continue on this theme, with a specific focus on air and water, by reading *A River Ran Wild: An Environmental History*, by Lynne Cherry. The book tells the true story of the Nashua River in New England. The first people to encounter the bounty of the river are the Algonquin speaking Nashua people. Over time, as European settlers came and industrialization followed, the river became seriously polluted. A descendant of the Nashua started a movement to return the river to its pristine state.

Introduce the guided inquiry question: **How can we keep our water clean?**

Portage & Main Press, 2019 · *Hands-On Science for British Columbia · Land, Water, and Sky for Grades K–2* · ISBN: 978-1-55379-797-5

18

Explore Part One

As a class, read books about water pollution, discussing examples, problems, and solutions.

Title a sheet of chart paper "Water Pollution," and divide the chart into two columns. Label the columns "Sources of Water Pollution" and "Solutions." Ask:

- What causes water pollution around the world?

Record students' ideas in the first column of the chart. Ask:

- How can we stop or decrease water pollution?

Have students focus on each source of pollution and provide solutions for these problems. Record the solutions in the second column of the chart.

Explore Part Two

NOTE: As students conduct the following investigation, have them photograph the mixing of oil with water and the results they observe when oil covers feathers and faux fur.

Begin by discussing the importance of water to all living things. Ask:

- How do humans use water?
- How is water important to other living things?
- How does our water get polluted?
- How does polluted water affect humans?
- How does polluted water affect plants and other animals?

Display pictures and books about water pollution, focusing on oil spills. Explain that large oil tankers carry oil across the oceans to countries all over the world. Sometimes the tankers have accidents, and the oil spills into the water.

Display the vegetable oil for students to examine. Allow them touch the oil and rub it on their fingers to see how it feels. Ask:

- What do you think would happen if an oil tanker hit a rock or experienced bad weather and spilled oil into the ocean?

Fill a 2-litre pop bottle with one litre of water. Add a few drops of food colouring to the water. Ask:

- What do you think will happen if oil is added to the water?

Now, have students test their predictions. Add 500 mL of vegetable oil to the 2-litre bottle. Put the cap on, then shake and rock the bottle to make waves. Have students observe what happens to the oil. Ask:

- Does the oil mix with the water?
- What would oil look like on the ocean?
- What do you think would happen to animals that lived in an ocean filled with oil?
- What would happen to birds that lived along the shoreline of that same ocean?

Next, pour some of the oil and water mixture into a basin or dishpan. Then provide each student with a feather. Ask:

- How are feathers important to a bird? (they provide warmth and are needed for flying)
- What do you think would happen if a bird got oil on its feathers?

Have students dip the feathers into the oil and observe what happens. Ask:

- If a bird's feathers were covered with oil, do you think the bird could survive?

NOTE: After an oil spill, oil soaks into the bird's feathers, so it cannot fly. Also, its feathers no longer provide insulation to help the bird stay warm. Oil also gets into the eyes and mouths of the bird, which can blind and poison it.

Portage & Main Press, 2019 · Hands-On Science for British Columbia · Land, Water, and Sky for Grades K–2 · ISBN: 978-1-55379-797-5

Provide each student with a small piece of faux fur. Ask:

■ How do you think an oil spill would affect a mammal living near an ocean?

Repeat the investigation with the faux fur. Have students dip the fur into the oil and observe the effects. Discuss the impact this would have on animals.

Now, challenge students to suggest ways the oil spill could be cleaned up. Record their ideas on chart paper, and help them collect materials they suggest (e.g., paper towels, cotton batting). Provide plenty of time for students to experiment with cleaning up the oil spill. Encourage students to be creative in developing novel ways to clean up.

Expand

Provide students with an opportunity to explore water pollution further by posing their own questions for individualized inquiry. They may wish to:

■ Initiate a project at the Makerspace, such as designing and constructing a device to clean water.

■ Explore Loose Parts bins related to water pollution with a variety of bird feathers for students to examine, discuss, sort, and measure, and to inspire further investigations. Also provide a bin of plastic animals for role playing.

■ Conduct research into local water pollution issues.

■ Write an MP or MLA about local water pollution concerns. Local MPs and MLAs may respond to student inquiries, inspiring student voice and action. It can also help students to have an authentic task to consider when researching water pollution issues.

■ Create a picture book about water pollution and solutions.

■ Explore pollution at the water table. Ensure there are plastic animals for role playing, as well as feathers and faux fur.

■ Conduct an investigation or experiment based on their own inquiry questions.

As students explore and select ideas to expand learning, provide support and guidance as needed, and offer access to materials and resources that will enable students to conduct their chosen investigations.

Formative Assessment

Conference with students individually. Have them reflect on what they have learned about ways to keep water clean. Encourage them to provide specific examples of things they can do at home, school, and in their community. This focuses on generating and introducing new or refined ideas when problem solving. Use the INDIVIDUAL STUDENT OBSERVATIONS template, on page 51, to record results. Provide descriptive feedback to students about how they used new ideas to solve this problem.

Embed Part One: Sharing Circle

Revisit the guided inquiry question: **How can we keep our water clean?** Have students share their experiences and knowledge, provide examples, and ask further inquiry questions.

Embed Part Two

■ Add to the concept web as students learn new concepts, answer some of their own inquiry questions, and ask new inquiry questions.

■ Add new terms and illustrations to the class word wall. Include the words in languages other than English, as appropriate.

■ Focus on students' use of the Core Competencies. Have students reflect on how they used one of the Core Competencies

▶

Portage & Main Press, 2019 · Hands-On Science for British Columbia · Land, Water, and Sky for Grades K–2 · ISBN: 978-1-55379-797-5

Portage & Main Press, 2019 · Hands-On Science for British Columbia · Land, Water, and Sky for Grades K–2 · ISBN: 978-1-55379-797-5

18

(Thinking, Communicating, or Personal and Social Skills) during the various lesson activities. Project one of the CORE COMPETENCY DISCUSSION PROMPTS templates (pages 38–42), and use it to inspire group reflection. Referring to the template, choose one or two "I Can" statements on which to focus. Students then use the "I Can" statements to provide evidence of how they demonstrated that competency. Ask questions directly related to that competency to inspire discussion. For example:

- What are you proud of in your learning today? (Personal Awareness and Responsibility)

Have students reflect orally, encouraging participation, questions, and the sharing of evidence. (See page 29 for more information on these templates.)

As part of this process, students can also set goals. For example, ask:

- What would you do differently next time and why?
- How will you know if you are successful in meeting your goal?

- To encourage self-reflection, provide prompts that students can use to cite examples of how they have used the Core Competencies in their learning. For this purpose, the CORE COMPETENCY SELF-REFLECTION FRAMES (pages 43–47) can be used throughout the learning process. There are five frames provided to address the Core Competencies: Communication, Creative Thinking, Critical Thinking, Positive Personal and Cultural Identity, and Personal Awareness and Responsibility. Teachers can conference individually with students to support self-reflection, or students may complete prompts using words and pictures.

Again, have students set goals by considering what they might do differently on future tasks and how they will know if they are successful in meeting their goal.

NOTE: Use the same prompts from these templates over time to see how thinking changes with different activities.

Enhance

- **Family Connections**: Provide students with the following sentence starter:
  - As a family, we can keep our local bodies of water clean by _____.

  Have students complete the sentence starter at home. Family members can help students draw and write about this topic. Have students share their sentences with the class.

- If students become passionate about water pollution, they may wish to plan presentations for assemblies, parents, or community groups; make posters, book marks, or class books on the topic.

- Read *Michael Bird-Boy* by Tomie dePaola to the class. This is a story about a young boy who discovers a factory polluting his community. Michael sets out to find solutions to this problem. Discuss the sequence of events in the story, the problem, and solution. Have students write letters to local factories or manufacturing plants to find out how they are controlling pollution.

- Challenge students to find ways to clean the feather and faux fur so animals could survive the oil spill.

- Show students videos about how oil spills have been cleaned up and how animals are rescued during these types of disasters.

- Discuss, and have students research, local and regional water issues, such as:
  - Trans Canada Pipeline and concerns regarding potential oil spills
  - increased tanker traffic near ports
  - Site C dam

- unsafe lead content in school water fountains (this may hit too close to home for some teachers) go to <https://vancouversun.com/news/local-news/more-than-half-of-b-c-s-school-districts-had-unsafe-lead-levels-in-drinking-water-sources-in-2016>
- Nestle bottling tap water in British Columbia
- fracking and LNG projects

■ Conduct the "Magic Water" experiment with students. Place four clear plastic cups onto a tray. Fill the cups halfway with water, and add a different colour of food colouring to each cup. Pour 75 mL of cooking oil on top of the coloured water in each cup.

Have students predict what colour a 2 x 4 cm strip of white construction paper will be after it is dipped into one of the cups. Then, have students test their predictions by dipping the paper into the selected cup. Have them repeat the process with the other cups.

■ Tell students to observe carefully and explain why the strips of paper do not turn the same colour as the water into which they are dipped.

NOTE: The oil floating on top of the coloured water coats the strips of paper as they are dipped into each cup. The coloured water cannot adhere to the oil, and, therefore, does not stain the paper. This will further illustrate for students how oil from an oil spill affects the feathers or fur of animals.

Portage & Main Press, 2019 · *Hands-On Science for British Columbia · Land, Water, and Sky for Grades K–2* · ISBN: 978-1-55379-797-5

Inquiry Project: How Does Extreme Weather Impact the Land, Water, and Sky?

19

Information for Teachers

Students will demonstrate their understanding of the relationship between land, water, sky, and extreme weather, as they role-play as meteorologists.

A *meteorologist* analyzes and forecasts the weather, provides consultation on atmospheric phenomena, and conducts research into the processes and phenomena of weather, climate, and atmosphere.

A *weather reporter* broadcasts weather forecasts. They must communicate well.

Extreme weather is weather on a larger, more serious, and devastating scale. These events are rare; only five percent (or less) of weather can be qualified as extreme weather. Examples include:

Blizzard: a severe snowstorm characterized by strong sustained winds of at least 56 km/h and lasting for a prolonged period of time—typically three hours or more.

Cyclone: a system of winds that rotate inward to an area of low atmospheric pressure, circulating counterclockwise in the northern hemisphere and clockwise in the southern hemisphere.

Flood: an excess or overflow of water that submerges usually dry land.

Hurricane: a storm with violent wind that usually begins as an ordinary storm over a warm tropical ocean; the storm becomes bigger as it absorbs heat and moisture from the warm ocean water.

Monsoon: a seasonal prevailing wind in South and Southeast Asia, which either blows from the southwest, between May and September and brings rain (the wet monsoon), or from the northeast between October and April (the dry monsoon).

Sandstorm/dust storm: arises when a gust front or other strong wind blows loose sand and dirt from a dry surface; common in arid and semi-arid regions.

Tornado: violent, twisting, dark, funnel-shaped column of air that is wide at the top and narrow at the bottom.

Tsunami: a series of waves caused by the displacement of a large portion of a body of water, generally an ocean or a large lake.

Typhoon: a mature tropical cyclone that occurs in the China Sea.

Materials

- *Freddy the Frogcaster* by Janice Dean
- books about meteorology and extreme weather, and other resources related to weather
- chart paper
- student dictionaries
- poster paper
- markers
- computer/tablet with internet access (optional)
- variety of craft and musical materials (e.g., Plasticine, pipe cleaners, markers, paint, pictures, glue, foam, musical instruments)
- writing paper
- concept web (from lesson 3)

Engage

Read the book *Freddy the Frogcaster* by Janice Dean. Ask students:

- What are some words used to describe people who work with the weather?

On chart paper, record a list as students suggest terms (e.g., *forecaster*, *weather person*, *weather reporter*, *scientist*) .

Now, record the term *meteorologist* on chart paper. Ask:

Portage & Main Press, 2019 · *Hands-On Science for British Columbia · Land, Water, and Sky for Grades K–2* · ISBN: 978-1-55379-797-5

19

- What do you think this word means?

Have students check dictionaries and discuss the term. Determine a class definition for meteorologist.

Introduce the guided inquiry question: **How does extreme weather impact the land, water, and sky?**

Explore Part One

Explain to students they will have the opportunity to be a meteorologist for the day. To prepare for this job they will be researching an example of extreme weather.

Organize the class into working groups, and provide each group with several books and other resources about weather. Give each group a sheet of chart paper and a marker, and have them brainstorm a list of examples of extreme weather.

Then, have each group share their list with the rest of the class, and create one class list.

Now, explain to students that they will conduct individual inquiry projects about extreme weather. Each student will choose one example of extreme weather and then come up with three questions they would like to research about this type of weather. Students will then share their researched answers in one of the following ways:

- Write and present a weather report for a severe weather event.
- Use the research to accurately present the information as a reporter for the Weather Network would.
- Make a short video clip or screencast
- Write a song describing how to remain safe in severe weather conditions.
- Create a poster or collage about this type of extreme weather.
- Make a model of this type of extreme weather.

- Create a travel brochure for an area that is typically affected by severe weather.
- Present a demonstration that models the extreme weather example.

As a class, review each extreme weather example on the list. Have each student select one example of extreme weather to explore further, and record their name beside the example in the list. More than one student may research each type of extreme weather.

Brainstorm with students examples of questions they might ask. Using hurricane as an example:

- What is a hurricane?
- Where in the world is a hurricane most likely to occur?
- What causes a hurricane to form?
- What steps can be taken to prepare for a hurricane and to help keep people safe?
- What damage can occur from a hurricane?
- What other interesting information did you find about hurricanes?

As a class, co-construct criteria for students' inquiry projects. For example:

- creates three questions
- uses more than one source to gather information
- researches and answers all three questions
- plans a way to present their learning to the class

Have students identify and record their three inquiry questions, as well as the criteria for the project.

Explore Part Two

Invite a local meteorologist or weather reporter to present to the class about extreme weather conditions. Weather specialists can be accessed through Environment Canada, The Weather Network, and local news stations. This also

Portage & Main Press, 2019 · Hands-On Science for British Columbia · Land, Water, and Sky for Grades K–2 · ISBN: 978-1-55379-797-5

Portage & Main Press, 2019 · Hands-On Science for British Columbia · Land, Water, and Sky for Grades K–2 · ISBN: 978-1-55379-797-5

19

provides an opportunity for students to share their own experiences with extreme weather.

NOTE: Be sensitive to students' past experiences, as some may have difficulty recalling traumatic events related to extreme weather conditions.

Also invite local Elders and Knowledge Keepers to share information and stories related to extreme weather.

NOTE: See Indigenous Perspectives and Knowledge, page 9, for guidelines for inviting Elders and Knowledge Keepers to speak to students.

With all guest speakers, have students brainstorm and record questions ahead of time, in order to gather information for their inquiry projects.

Explore Part Three

Provide students with a variety of resources to use for their inquiry projects. Also, encourage them to seek out information on their own. Model for students how to look for information in a nonfiction text. They may use pictures and text, as well as peer/adult support for the research process.

Guide and model the process of transferring students' researched information to a presentation format. As with the research process, students will require time and support to create their presentations/projects.

Give students an opportunity to share their learning by presenting their final projects to the class.

Formative Assessment

Record the co-constructed criteria for the inquiry project on the RUBRIC, on page 53. When students present their inquiry projects, assess and record results. The focus herein is on representing observations and ideas about land, water, and sky. Provide descriptive feedback to

students about how they shared observations and ideas.

Student Self-Assessment

Have students reflect on their learning by completing the STUDENT REFLECTION template, on page 36. Encourage them to consider the criteria for their inquiry projects, as well as their learning throughout the module. Focus here on demonstrating curiosity and a sense of wonder about land, water, and sky.

Embed Part One: Sharing Circle

Revisit the guided inquiry question: **How does extreme weather impact the land, water, and sky?** Have students share their knowledge and provide examples to consolidate learning.

Embed Part Two

- Add to the concept web as students learn new concepts and answer some of their own inquiry questions.
- Add new terms and illustrations to the class word wall. Include the words in other languages, as appropriate.
- Focus on students' use of the Core Competencies. Have students reflect on how they used one of the Core Competencies (Thinking, Communicating, or Personal and Social Skills) during the various lesson activities. Project one of the CORE COMPETENCY DISCUSSION PROMPTS templates (pages 38–42), and use it to inspire group reflection. Referring to the template, choose one or two "I Can" statements on which to focus. Students then use the "I Can" statements to provide evidence of how they demonstrated that competency. Ask questions directly related to that competency to inspire discussion. For example:
  - How did you show that you were an active listener today? (Communication)

Have students reflect orally, encouraging participation, questions, and the sharing of evidence. (See page 29 for more information on these templates.)

As part of this process, students can also set goals. For example, ask:

■ What would you do differently next time and why?

■ How will you know if you are successful in meeting your goal?

■ To encourage self-reflection, provide prompts that students can use to cite examples of how they have used the Core Competencies in their learning. For this purpose, the CORE COMPETENCY SELF-REFLECTION FRAMES (pages 43–47) can be used throughout the learning process. There are five frames provided to address the Core Competencies: Communication, Creative Thinking, Critical Thinking, Positive Personal and Cultural Identity, and Personal Awareness and Responsibility. Teachers can conference individually with students to support self-reflection, or students may complete prompts using words and pictures. Again, have students set goals by considering what they might do differently on future tasks and how they will know if they are successful in meeting their goal.

NOTE: Use the same prompts from these templates over time to see how thinking changes with different activities.

Enhance

■ **Family Connections**: Have students take their projects home to present to their families. Also provide the following sentence starter:

■ I taught my family about _____. They learned _____.

Have students complete the sentence starter at home. Family members can help students draw and write about this topic. Have students share their sentences with the class.

■ Discuss and research global warming and its effects on weather.

Unit Assessment Summary

■ Consider having a collection of student work gathered in a portfolio, so students can examine and discuss these artifacts of learning during a conference. This collection may include photographs they have taken, drawings, place-based journals, and so on. This will allow students to recall specific activities and learning experiences and to reflect on their use of the Core Competencies throughout this module.

■ Have students take home a copy of the FAMILY AND COMMUNITY CONNECTIONS ASSESSMENT template on page 57. Have them complete the sheet with a family or community member (with permission) to reflect on their learning about the features of the land, water, and sky.

■ Have students focus on the CORE COMPETENCIES STUDENT REFLECTIONS UNIT SUMMARY template, on page 48, to reflect on their use of the Core Competencies throughout the module. Students' reflections are recorded in the rectangle on the template (using pictures and text). Next, the student considers next steps in learning as related to that particular Core Competency. These reflections are recorded in the arrow in the template, again, using words and drawings.

■ Review all assessment templates completed throughout the module. This includes all documentation for student self-assessment, formative assessment, and summative assessment.

▶

Portage & Main Press, 2019 · *Hands-On Science for British Columbia · Land, Water, and Sky for Grades K–2* · ISBN: 978-1-55379-797-5

Appendix: Image Banks

Images in this appendix are thumbnails from the Image Banks referenced in the lessons. Corresponding full-page, high-resolution images can be printed or projected for the related lessons, and are found on the Portage & Main Press website at: <www.portageandmainpress.com/product/HOSLANDWATERSKYK2/>. Use the password **WATERCYCLE** to access the download for free. This link and password can also be used to access the reproducible templates for this module.

Please follow these steps to retrieve the images and reproducible templates for this book.

1. Go to <www.portageandmainpress.com/product/HOSLANDWATERSKYK2/>.
2. Type the password **WATERCYCLE** into the password field.
3. Select Add to Cart.
4. Select View Cart.
5. Select Proceed to Checkout. No coupon code is required.
6. Enter your billing information or log in to your existing account using the prompt at the top of the page.
7. Select Place Order.
8. Under Order Details, click the link for your download.
9. Save the file to the desired location on your computer.

NOTE: This is a large file. Download times will vary due to your internet speeds.

Lesson 8: What Do We Know About Seasonal Changes?
Seasons

1. Skiing (Winter)

2. Snowman (Winter)

3. Shovelling Snow (Winter)

4. Snowfall (Winter)

5. Snowshoeing (Winter)

6. Flower in Bloom (Spring)

7. Puddle Jumping (Spring)

8. Cherry Tree (Spring)

9. Pine Tree (Spring)

10. Bees Collecting Pollen (Summer)

11. Hummingbird (Summer)

12. Garden (Summer)

Portage & Main Press, 2019 · *Hands-On Science for British Columbia · Land, Water, and Sky for Grades K–2* · ISBN: 978-1-55379-797-5

13. Kayaking (Summer)

14. Hiking (Summer)

15. Children Playing in Sunshine (Summer)

16. Swimming (Summer)

17. Tree With Colourful Leaves (Autumn)

18. Fleetwood Elementary School (Autumn)

19. Carved Pumpkins (Autumn)

20. Corn Maze (Autumn)

21. Hayride (Autumn)

Image Credits:

1 – Ski kids by Ruth Hartnup. Used under CC by 2.0 licence.

2 – Snowman sitting by Ruth Hartnup. Used under CC by 2.0 licence.

3 – Shovelling the neighbours' path by Ruth Hartnup. Used under CC by 2.0 licence.

4 – Yaletown Park Snowfall by Robin Monks. Used under CC by 2.0 licence.

5 – Snowshoeing at Cypress by Christine Rondeau. Used under CC by 2.0 licence.

6 – Raindrops on snowdrops by Ruth Hartnup. Used under CC by 2.0 licence.

7 – puddle jumping by Nick&Shelien Hadfield. Used under CC by 2.0 licence.

8 – 20170322-APHIS-LSC-0259 by Lance Cheung. Courtesy of the U.S. Department of Agriculture. Used under CC by 2.0 licence.

9 – Pine Tree Bud by George Donnelly. Used under CC by 2.0 licence.

10 – Bee Meeting by John Anderson. Used under CC by 2.0 licence.

11 – Rufous Hummingbird by Ed Dunens. Used under CC by 2.0 licence.

12 – Shoot the Shooter by Sheila Sund. Used under CC by 2.0 licence.

13 – Kayak by Ed Dunens. Used under CC by 2.0 licence.

14 – Untitled by Edna Winti. Used under CC by 2.0 licence.

15 – Playing nicely together by Ruth Hartnup. Used under CC by 2.0 licence.

16 – Shallow waters at Locarno Beach by Ruth Hartnup. Used under CC by 2.0 licence.

17 – Cherry tree autumn grove by Ruth Hartnup. Used under CC by 2.0 licence.

18 – 130818-13 by waferboard. Used under CC by 2.0 licence.

19 – Four Halloween pumpkins by Ruth Hartnup. Used under CC by 2.0 licence.

20 – 20100822_029.jpg by Roland Tanglao. Used under CC0 licence.

21 – AppleBarn Farm by Marcin Chady. Used under CC by 2.0 licence.

Portage & Main Press, 2019 · Hands-On Science for British Columbia · Land, Water, and Sky for Grades K–2 · ISBN: 978-1-55379-797-5

Lesson 9: How Do Seasonal Changes Affect Plants?
Indigenous Plants Throughout the Seasons

1. Western Redcedar (Summer)

2. Acorns (Autumn)

3. Blueberries (Summer)

4. Sage (Spring)

5. Pacific Crab Apple (Autumn)

6. Hazelnuts (Autumn)

7. Mustard Plant (Summer)

8. Nodding Wild Onion (Summer)

9. Watercress (Spring)

10. Vine Maple Tree (Autumn)

11. Fir Tree (Spring)

12. Tamaracks (Autumn)

Portage & Main Press, 2019 · Hands-On Science for British Columbia · Land, Water, and Sky for Grades K–2 · ISBN: 978-1-55379-797-5

13. Evergreen Tree (Winter)

14. Willow Tree (Summer)

15. Birch Trees (Winter)

Image Credits:

1 – World's Largest Western Red Cedar by John Williams. Used under CC by 2.0 licence.

2 – Acorns by JB Kilpatrick. Used under CC by 2.0 licence.

3 – Blueberry bushes by Ruth Hartnup. Used under CC by 2.0 licence.

4 – Sage by Isaac Wedin. Used under CC by 2.0 licence.

5 – Ripening fruits of Pacific crab apple, Island View Beach, Saanich Peninsula, British Columbia by Dr. Richard J. Hebda.

6 – Hazenut, Hazel, Nut, Plant, Brown, Food, Autumn, Open by Wolfgang Eckert. Used under CC0 licence.

7 – Garin Regional 20150204 by Sarah Sammis. Used under CC by 2.0 licence.

8 – Nodding Onion, Allium Cernuum by Karuna Poole. Used under CC0 licence.

9 – Watercress, Blossom, Bloom, Spring by Alois Grundner. Used under CC0 licence.

10 – H20091119-1793--Acer circinatum -- RPBG by John Rusk. Used under CC by 2.0 licence.

11 – Fir tree in spring by Kam Abbott. Used under CC by 2.0 licence.

12 – Mixed Silviculture by Gerry. Used under CC by 2.0 licence.

13 – Frosted Boughs by Luke Jones. Used under CC by 2.0 licence.

14 – Willow_1 by Andrew Taylor. Used under CC by 2.0 licence.

15 – Birch Trees by Kathleen Franklin. Used under CC by 2.0 licence.

Lesson 10: How Do Seasonal Changes Affect Animals?
Animals of British Columbia

1. Sockeye Salmon (Summer)

2. Columbia Spotted Frog (Summer)

3. Red-Eared Slider Turtles (Spring)

4. Bald Eagle (Winter)

5. Grizzly Bears (Spring)

6. Grey Wolf (Winter)

Portage & Main Press, 2019 · Hands-On Science for British Columba · Land, Water, and Sky for Grades K–2 · ISBN: 978-1-55379-797-5

7. Bighorn Sheep (Summer)

8. Grey Whale (Summer)

9. Humpback Whales (Spring)

10. Killer Whales (Summer)

11. Mountain Goats (Spring)

12. Walleye (Summer)

13. Pacific White-Sided Dolphin (Summer)

14. Sand Crab (Summer)

Image Credits:

1 – July 2010, Spawning male sockeye by Thomas Quinn, courtesy of USEPA Environmental-Protection-Agency. Used under CC0 licence.

2 – Columbia spotted frog by Aubree Benson. Courtesy of the U.S Forest Service.

3 – soaking up the sun by Ruth Hartnup. Used under CC by 2.0 licence.

4 – Bald Eagle by @herewasthere. Used under CC by 2.0 licence.

5 – Bear-with-3-spring-cubs-6-19-2014_2 by Katmai National Park and Preserve. Used under CC by 2.0 licence.

6 – Gray Wolf by Lucas. Used under CC by 2.0 licence.

7 – Curious by David Meurin. Used under CC by 2.0 licence.

8 – anim1723 by Dr. Steven Swartz, NOAA/NMFS/OPR. via NOAA Photo Library. Used under CC by 2.0 licence.

9 – Whales by Ed Lyman/NOAA, under NOAA Permit #14682. This image is in the public domain.

10 – Born Free by Christopher Michel. Used under CC by 2.0 licence.

11 – Mother and child mountain goats together by David Fulmer. Used under CC by 2.0 licence.

12 – Walleye by U.S. Fish and Wildlife Service, Eric Engbretson via the National Digital Library, United States Fish and Wildlife Service. Used under Public Domain Mark 1.0 licence.

13 – A Pacific White-Sided Dolphin by Tom Keickhefer (NOAA) by Tom Keickhefer, NOAA via pingnews. Used under Public Domain Mark 1.0 licence.

14 – Not a hermit crab! by hobvias sudoneighm. Used under CC by 2.0 licence.

Portage & Main Press, 2019 · Hands-On Science for British Columbia · Land, Water, and Sky for Grades K–2 · ISBN: 978-1-55379-797-5

Lesson 12: Which Objects Do We See in the Daytime Sky?
Rainbows

1. Urban Rainbow

2. Rural Rainbow

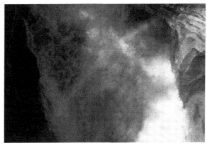

3. Rainbow Over Water

4. Close-Up of Rainbow

5. Rainbow Under Cloud Shade

6. Double Rainbow

7. Double Rainbow

Image Credits:

1 – Sunrise rainbow by Ruth Hartnup. Used under CC by 2.0 licence.

2 – Malahat, B.C. by Don Dykun. Used under CC by 2.0 licence.

3 – Looking down at rainbows over Sims Creek's Canyon by McKay Savage. Used under CC by 2.0 licence.

4 – Victoria 20141227 by Sarah Sammis. Used under CC by 2.0 licence.

5 – 2011 11 Remembrance Day by Blake Handley. Used under CC by 2.0 licence.

6 – double rainbow by waferboard. Used under CC by 2.0 licence.

7 – rainbow McBride July 2013 by Jim Swanson. Used under CC by 2.0 licence.

Lesson 13: Which Objects Do We See in the Nighttime Sky?
The Moon

1. Solar Eclipse

2. New Moon

3. Waxing Crescent

Portage & Main Press, 2019 · Hands-On Science for British Columbia · Land, Water, and Sky for Grades K–2 · ISBN: 978-1-55379-797-5

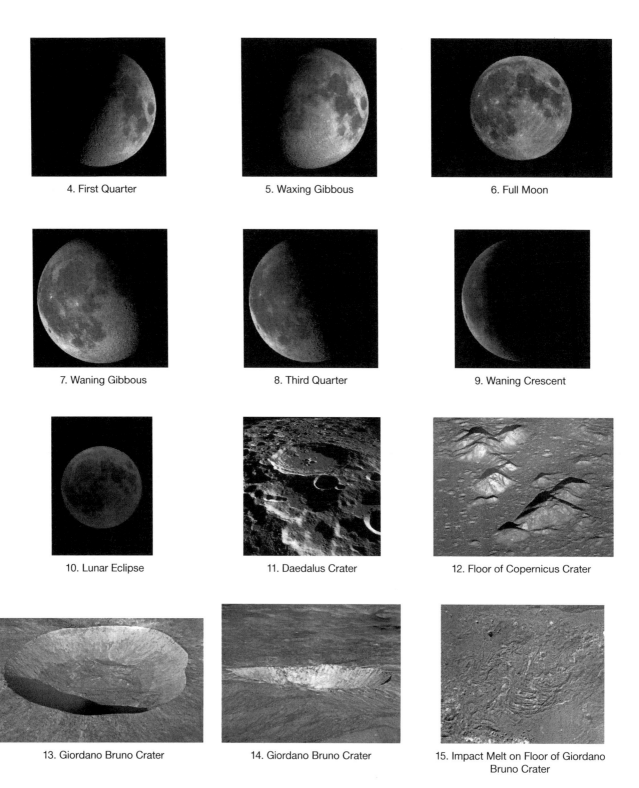

4. First Quarter

5. Waxing Gibbous

6. Full Moon

7. Waning Gibbous

8. Third Quarter

9. Waning Crescent

10. Lunar Eclipse

11. Daedalus Crater

12. Floor of Copernicus Crater

13. Giordano Bruno Crater

14. Giordano Bruno Crater

15. Impact Melt on Floor of Giordano Bruno Crater

Portage & Main Press, 2019 · Hands-On Science for British Columbia · Land, Water, and Sky for Grades K–2 · ISBN: 978-1-55379-797-5

Image Credits:

1 – Total Solar Eclipse 2017 by Bernd Thaller. Used under CC by 2.0 licence.

2 – New Moon by NASA/Goddard Space Flight Center Scientific Visualization Studio. Used under CC by 2.0 licence.

3 – Waxing crescent by NASA/Goddard Space Flight Center Scientific Visualization Studio. Used under CC by 2.0 licence.

4 – First Quarter by NASA/Goddard Space Flight Center Scientific Visualization Studio. Used under CC by 2.0 licence.

5 – Waxing Gibbous by NASA/Goddard Space Flight Center Scientific Visualization Studio. Used under CC by 2.0 licence.

6 – Full Moon by Javier Morales. Used under CC by 2.0 licence.

7 – Waning Gibbous by NASA/Goddard Space Flight Center Scientific Visualization Studio. Used under CC by 2.0 licence.

8 – Third Quarter by NASA/Goddard Space Flight Center Scientific Visualization Studio. Used under CC by 2.0 licence.

9 – Waning Crescent by NASA/Goddard Space Flight Center Scientific Visualization Studio. Used under CC by 2.0 licence.

10 – Lunar Total Eclipse on July 27, 2018 (100_2006) by Giuseppe Donatiello. Used under CC0 licence.

11 – as11_44_6609 by the National Museum of the U.S. Navy. Used under Public Domain Mark 1.0 licence.

12 – Floor of Copernicus Crater (Moon) 2 by the Lunar Reconnaissance Orbiter Camera. Used under CC0 licence.

13 – Giordano Bruno Crater (Moon) 1 by the Lunar Reconnaissance Orbiter Camera. Used under CC0 licence.

14 – Giordano Bruno Crater (Moon) 2 by the Lunar Reconnaissance Orbiter Camera. Used under CC0 licence.

15 – Impact melt in Giordano Bruno Crater (Moon) by the Lunar Reconnaissance Orbiter Camera. Used under CC0 licence.

Lesson 14: What Forms of Water Are Found on Earth?

Forms of Water on Earth

1. Skeena and Bulkley Rivers

Location: 'Ksan, British Columbia.

2. Moose Lake

Location: Mount Robson Provincial Park, British Columbia.

3. Stream

Location: Salt Spring Island, British Columbia.

4. Pond

Location: Vancouver, British Columbia.

5. Marsh

Location: Finn Slough, British Columbia.

6. The North Pacific Ocean

Location: Chesterman Beach in Tofino, British Columbia.

Portage & Main Press, 2019 · *Hands-On Science for British Columbia · Land, Water, and Sky for Grades K–2* · ISBN: 978-1-55379-797-5

7. Puddle

8. Wapta Falls
Location: Field, British Columbia.

9. Wapta Falls
Location: Field, British Columbia.

10. Rain

11. Rain

12. Snow

13. Snow

14. Sleet

15. Hail

16. Hail

17. Dew

18. Droplet

Portage & Main Press, 2019 · Hands-On Science for British Columbia · Land, Water, and Sky for Grades K–2 · ISBN: 978-1-55379-797-5

Image Credits:

1 – 'Ksan at Skeena/Bulkley confluence by Sam Beebe. Used under CC by 2.0 licence.

2 – IMG_3122 by Andy M Smith. Used under CC by 2.0 licence.

3 – Broad Leaf Maple Trees, Salt Spring island, BC, Canada by Steven Petty. Used under CC by 2.0 licence.

4 – Pond by Jennifer Copley. Used under CC by 2.0 licence.

5 – Finn Slough by dvdmnk. Used under CC by 2.0 licence.

6 – Chesterman Beach by Ruth Hartnup. Used under CC by 2.0 licence.

7 – Untitled by eflon. Used under CC by 2.0 licence.

8 – Wapta Falls by Travis Purcell. Used under CC by 2.0 licence.

9 – DSC00491 by Travis Purcell. Used under CC by 2.0 licence.

10 – trash island by velcr0. Used under CC by 2.0 licence.

11 – Mountains and rain showers by Ruth Hartnup. Used under CC by 2.0 licence.

12 – Crazy Snow by Alison and Fil. Used under CC by 2.0 licence.

13 – 2015 RVA Macro Snow by Eli Christman. Used under CC by 2.0 licence.

14 – Snow falls in Whistler Creekside by Ruth Hartnup. Used under CC by 2.0 licence.

15 – Hailstorm on #5 green, Skedhult by Jonas Löwgren. Used under CC by 2.0 licence.

16 – Hail in May by Ruth Hartnup. Used under CC by 2.0 licence.

17 – Macro dew drops by Faruk Ateş. Used under CC by 2.0 licence.

18 – Codevasf 43 anos by Codevasf. Used under CC by 2.0 licence.

Lesson 15: How Does Water Move Through the Water Cycle?
Mountains

1. Mount Seymour, British Columbia

2. Mount Robson, British Columbia

3. Mount Assiniboine, British Columbia

4. Grouse Mountain, British Columbia

5. Victoria Peak, British Columbia

6. Mount Grant, British Columbia

Image Credits:

1 – Snowy Vancouver from Queen Elizabeth Park by Ruth Hartnup. Used under CC by 2.0 licence.

2 – O Canada by Jeff Pang. Use under CC by 2.0 licence.

3 – Crisp by Jeff Pang. Used under CC by 2.0 licence.

4 – Jericho Beach by Jacek S. Used under CC by 2.0 licence.

5 – Cloudy Victoria Peak at Lake Louise by Laszlo Ilyes. Used under CC by 2.0 licence.

6 – Mount Grant by David Meurin. Used under CC by 2.0 licence.

Creative Commons Licences

CC BY 2.0: Attribution 2.0 Generic <https://creativecommons.org/licences/by/2.0/>

CC BY-SA 2.0: Attribution-ShareAlike 2.0 Generic <https://creativecommons.org/licenses/by-sa/2.0/>

CC BY 4.0: Attribution 4.0 International <https://creativecommons.org/licences/by/4.0/>

Public Domain Mark 1.0: <https://creativecommons.org/publicdomain/mark/1.0/>

CC0 1.0 Universal: <https://creativecommons.org/publicdomain/zero/1.0/>

About the Contributors

Jennifer Lawson, PhD, is the originator and senior author of the Hands-On series in all subject areas. Jennifer is a former classroom teacher, resource/special education teacher, consultant, and principal. She continues to develop new Hands-On projects, and also serves as a school trustee for the St. James-Assiniboia School Division in Winnipeg, Manitoba.

Rosalind Poon has been a science teacher and Teacher Consultant for Assessment and Literacy with the Richmond School District for the past 18 years. In her current role, she works with school teams to plan and implement various aspects of the curriculum by collaborating with teams in professional inquiry groups on topics such as descriptive feedback, inquiry, assessment, and differentiation. Her passions include her family, dragon boating, cooking with the Instant Pot and making sure that all students have access to great hands-on science experiences.

Deidre Sagert specializes in early years education, and is currently working as the Early Years Support Teacher for the St. James-Assiniboia School Division. She brings 20 years of experience to her current role where she mentors early years teachers in incorporating play-based learning and inquiry into all subject areas. She is passionate about ensuring all students have access to a stimulating environment where they are engaged in hands on experiences and authentic learning. She enjoys spending time with her family in nature for rejuvenation and inspiration.

Melanie Nelson is from the In-SHUCK-ch and Stó:lō Nations, and has experience teaching kindergarten through grade 12, as well as adults in the Lower Mainland of British Columbia. She has taught in mainstream, adapted, modified, and alternate settings, at the classroom, whole school, and district levels. Trained as an educator in science, Melanie approaches Western science through an Indigenous worldview and with Indigenous ways of knowing. Her Master of Arts thesis explored the experience of Indigenous parents who have a child identified as having special needs in school, and she is currently completing a Doctor of Philosophy in School Psychology at the University of British Columbia.

Lisa Schwartz has been a Teacher Consultant for Assessment and Literacy with the Richmond School District for the past six years. As a consultant, Lisa facilitates professional learning with small groups and school staffs on topics such as the redesigned curriculum, Core Competencies, differentiation, inquiry, and assessment. She also works side by side with teachers co-planning, co-teaching and providing demonstration lessons to highlight quality, research-based instruction that supports all learners. Lisa is passionate about engagement, joyful learning and success for all students.

Hetxw'ms Gyetxw (Brett D. Huson) is from the Gitxsan Nation of the Northwest Interior of British Columbia, Canada. Growing up in this strong matrilineal society, Brett developed a passion for the culture, land, and politics of his people, and a desire to share their knowledge and stories. Brett has worked in the film and television industry, and has volunteered for such organizations as Ka Ni Kanichihk and Indigenous Music Manitoba. Brett is the author of the Mothers of Xsan series of children's books. The first book in the series is *The Sockeye Mother*, which won The Science Writers and Communicators Book Award.

Portage & Main Press, 2019 · Hands-On Science for British Columbia · Land, Water, and Sky for Grades K–2 · ISBN: 978-1-55379-797-5

GRADES
K-2
British Columbia

hands-on
science
An Inquiry Approach

ORDER THE FULL SET OF K–2 RESOURCES TODAY AT
PORTAGEANDMAINPRESS.COM

ALSO AVAILABLE FOR GRADES 3–5!

PORTAGE &
MAIN PRESS